NUMBERS THROUGH THE AGES

9/12

513.
5

Hertfordshire

04/07

9 NOV 2007

Please renew/return this item by the last date shown.

So that your telephone call is charged at local rate, please call the numbers as set out below:

	From Area codes 01923 or 020:	From the rest of Herts:
Renewals:	01923 471373	01438 737373
Enquiries:	01923 471333	01438 737333
Textphone:	01923 471599	01438 737599

L32   www.hertsdirect.org/librarycatalogue

Numbers Through the Ages

Edited by
Graham Flegg

Reader in the History of Mathematics
Open University

M

MACMILLAN

in association with

The Open University

First published 1989

Published by
MACMILLAN EDUCATION LTD
Houndmills, Basingstoke, Hampshire RG21 2XS
and London
in association with
THE OPEN UNIVERSITY
Walton Hall
Milton Keynes MK7 6AA

Distributed in the USA by
Sheridan House Inc.
145 Palisade Street
Dobbs Ferry
NY10522

Typeset by Footnote Graphics,
Warminster

Printed in Hong Kong

British Library Cataloguing in Publication Data
Numbers through the ages.
1. Number system & numeration. Origins
I. Flegg, Graham, *1924–*
513′.5
ISBN 0–333–49130–0
ISBN 0–333–49131–9 pbk

Contents

Acknowledgements

The author and publishers wish to thank the following who have kindly given permission for the use of copyright material: Bibliothèque Nationale, Paris, for the reproductions of an abacus of the Gerbert School, Latin 8663, Fol. 49, and a Roman hand abacus, Cabinet des Médailles; The British Library for the reproductions of finger numbers from a thirteenth-century Spanish codex, in Lucia Paccioli's *Summa de arithmetica, geometria proportioni et proportionalita*, 1494, and the first page of *Crafte of Nombrynge*; The British Museum for reproductions of part of a Babylonian clay tablet, Mayan day and month signs, Mayan numbers on stone fragment found in British Honduras, extract from the Rhind Papyrus, and an extract from the Mathematical Leather Roll; Cambridge University Press for illustrations from *Science and Civilisation in China Vol. 3* by J. Needham, 1970; Dover Publications for an illustration from *A Source Book in Mathematics Vol. 1* by D. E. Smith, 1959; Ferranti International for photograph of Mk. 1 computer, ref. no. G869, 1950; Fototeca Unione for a reproduction of the inscription on the *Colonna rostrata*, No. 1324; Hessische Landes-und-Hochschul Bibliothek, Darmstadt for an extract from the *Carmen de algorismo* of Alexandre de Villa Dei, 2640, Nr. 15, folio 204^r; History of Science Society for material from 'Fundamental Steps in the Development of Numeration' by Carl B. Boyer, *Isis*, 1944, 35, pp. 153–68; IBM UK Ltd for illustrations of the Jacquard loom and the IBM Automatic Sequence Controlled Calculator; Institut Royal du Patrimoine Artistique, Brussels, for a schematic drawing of the Ishango bone; Linden-Museum Stuttgart for the reproduction of a Philippine sword, Luzon, Inv. No. 35 311; The MIT Press for material from *Number Words und Number Symbols* by K. Menninger, 1969; National Museum of Finland for the reproduction of a Finnish tally stick; National Museum of Ireland for the reproduction of an Irish gravestone from Aglish with Ogham characters; Österreiche Nationalbibliothek, Vienna, for an extract from a twelfth-century German manuscript on algorism, Codex 275; Raetisches Museum, Chur, for the reproduction of two 'milk sticks', XI 402 and H 1972.345; Royal Astronomical Society for the reproduction of a woodcut from Reisch, *Magarita philosophica*, 1508, showing Pythagoras using an abacus; Trustees of the Science Museum for reproductions of a Chinese suan pan, Russian ščët, two views of Pascal's calculating machine and Leibniz's calculating machine, calculator by C. Thomas, Babbage's difference engine, Scheutz's difference machine, Hollerith horizontal sorting machine and Babbage's operation card; Society of Antiquaries of London for the reproduction of British exchequer tallies; Springer Verlag GmbH for material from 'The Ritual Origin of Counting', A. Seidenberg in *Archive for History of Exact Sciences*, 1962; Staatliche Museen zu Berlin for reproduction of the Sumerian clay tablet, VAT 12593; TAP Service, Athens, for reproduction of a Salamis tablet; Vandenhoeck and Ruprecht for the reproduction of an alpscheit, Japanese soroban and diagram showing the stages of use of a soroban, bundle of

Alpine number billets, a number stick of split bamboo and capital tesserae; and with MIT Press for reproductions of a Chinese bank draft, Chinese commercial price tag and Chinese numeral; Yale University, Sterling Memorial Library, for the reproduction of cuneiform text, Babylonian Collection, YBC 7289.

Preface

The material of this book originally formed part of the Open University course AM289 *History of Mathematics*.

One fundamental aspect of the history of mathematics deals with the evolution of the concept of *number*, with the representation of numbers by words and symbols, and with the basic methods of calculation which have developed from ancient times to the present day. This aspect is of considerable general interest, and is accessible to the general reader including people who are not mathematically inclined. Since the Open University course was designed for a wide spectrum of students, the topic of 'number' was chosen as a special area of study suitable for non-mathematicians taking the course as an alternative to the *calculus* – an area of historical study requiring considerable mathematical experience.

The original course units which have been edited to form the present work were written by Professor B. L. van der Waerden, Dr Menso Folkerts, Dr E. Neuenschwander, Dr S. H. Hollingdale, and the present Editor, who was also chairman and principal author of the course, and was responsible for the academic editing of the material written by other authors. The units were as follows:

N1 *Counting I: Primitive and More Developed Counting Systems* (van der Waerden and Flegg)

N2 *Counting II: Decimal Number Words, Tallies, and Knots* (van der Waerden and Flegg)

N3 *Written Numbers* (van der Waerden and Folkerts)

N4 *Written Fractions* (van der Waerden)

N5 *Methods of Calculation* (van der Waerden and Neuenschwander)

11 *Mathematics and Man* – four sections from *Part A* (Hollingdale).

The material from Unit 11 was not part of the 'numbers' option.

In the course of the present editing, exercises and specimen answers have been deleted or have been incorporated into the main text. Statements of objectives and a number of inessential illustrations have also been removed. In some cases the order of presentation has been changed, and in a few instances additional material has been provided by the Editor for completeness or to reflect aspects of recent research.

AM 289 *History of Mathematics* was successfully presented by the Open University for ten years, the final year of presentation being 1985. Since that year there have been many requests received to have aspects of the course made once again available. In presenting this present work it is very much hoped that such requests may be, at least to a significant extent, adequately met.

Particular thanks are due to Professor van der Waerden for his co-operation in the preparation of the original 'numbers' units, and also for readily agreeing to the re-editing of his material to form this book.

The Open University   GRAHAM FLEGG
June 1988

1 Introduction

Number plays an essential role in our culture as indeed it does in any recognisable form of society. However far we go back in history, we can be certain that number has played its part in ways of thought and in human reaction to the world. This book is essentially about *number* in some of its many aspects.

However, the concern here is not with the abstract mathematical properties associated with number but with the practical problems of counting, writing numbers and performing simple arithmetical operations. The book begins by considering the counting systems which exist or have existed at various times and places. Counting systems are directly associated with the spoken word and sometimes with gestures of parts of the body. Discussion of the number words of languages both familiar and unfamiliar is followed by discussion of relationships between languages, particularly those of the Indo-European family to which English belongs. Subsequently the history of our own familiar decimal system of counting is considered together with rudimentary methods of recording numbers. The book continues with a survey of the various ways of writing numerals in ancient times and more recently in many parts of the world. Again, of special concern here is the history of the decimal system for writing numbers (including fractions): its origin and its development to the present day.

Number has many practical everyday applications which involve arithmetical calculation. In our familiar decimal system such calculation is comparatively easy. With other systems for writing numbers it has often not been so easy and early recourse was made to various aids to computation, one of the earliest of which was the counting board or abacus. The history of such aids is discussed in the final chapter.

The examination of the various aspects of number undertaken here makes no claim to be exhaustive. Essentially this work can do little more than lay the basis for a serious study of the cultural aspects of number. From time to time extracts are quoted from papers and books which treat the subject matter at greater depth. Readers who are stimulated to pursue the subject beyond the confines of this book are strongly recommended to study such works in their entirety.[1]

Counting systems

Chapters 2 and 3 are concerned with counting systems and number words. In addition to our familiar decimal counting system, several other counting methods are considered which were and (in some cases) are still used in parts of Africa, Asia, Australia and the Americas. Such methods include those known as *2-count*, *4-count*, *5–10 count*, *5–20 count*, and *counting on fingers and toes*. There is also some discussion of the Babylonian *10–60-system*. However, discussion is restricted primarily to the more important and characteristic systems.

The history of counting systems and number words is one of the most interesting parts of the history of civilisation. It is interesting not only in itself but also because of its close connection with the history of mathematics and the history of science in general. A good counting system, which enables us to express large numbers in spoken number words or written numerals, is an essential prerequisite for arithmetic, algebra, astronomy, physics and many other studies.

As far as possible not only the present state of various counting systems but also their history is investigated. Our decimal counting system and English number words can be traced over a period of nearly five thousand years. This is possible because the English language is akin to more ancient Germanic languages, such as Gothic, which in turn belong to the same great language family as Latin, Greek, Old Persian and Sanskrit. The languages of this family are called *Indo-European languages* because the area in which they were and still are spoken extends from India to Europe. All these languages are supposed to be derived from one hypothetical 'Original Indo-European language', which already existed in the period between 3000 and 2500 BC. In this original language number words up to 'hundred' (or perhaps 'thousand') existed and the method of forming composite number words like 'eight hundred and forty-five' was just the same as in our own language. All this is explained in Chapter 3.

However, on turning to examine the counting systems of primitive tribes, one's knowledge is limited to comparatively recent states. In exceptional cases, such as the Mayan and Aztec systems, one can go back one or two millennia but not more. In order to obtain information about the history of primitive counting methods, such as *counting by pairs*, hypotheses have to be made and these are always uncertain.

When finding similarities between the counting systems of tribes living in different countries or even in different continents, there are always two possible assumptions that can be made: a *diffusion* hypothesis (i.e. spreading from one country or continent to another) or an *independent invention* hypothesis (i.e. separate inventions of the same principle in many different places). In certain cases the diffusion hypothesis can be proved. For example, when indigenous tribes in South America are found to be using Spanish number words or when the non-white population of the United States is found to be speaking English and using the decimal counting system, clear cases of diffusion can be established.

However, in other cases where there is no direct historical evidence to support the diffusion hypothesis, the majority of scholars working in this field tend to assume independent invention. There is one notable exception to this rule. A. Seidenberg[2] puts forward the hypothesis that every one of the known counting systems was invented just once in one geographical location and then spread from this place of invention to those places where it is in use today.

How should one approach the opposing hypotheses of 'diffusion' and 'independent invention'? Should one seek to be entirely convinced by one of them, or should one take up a position somewhere between the two? Of course, the reader cannot be expected to decide at this point because as yet the arguments for and against each hypothesis have not been studied, nor has the evidence in particular cases. However, in the course of the book, the reader may well come to believe that only those arguments which favour the diffusion hypothesis are being presented and therefore insufficient justice is being done to the alternative hypothesis of independent invention which, at first sight, may seem the more plausible of the two. In fact, it is extremely difficult to find soundly based factually supported arguments in favour of independent invention. When assuming independent invention, scholars just present the argument that 'diffusion has not been conclusively proved', even though it is clear that this in itself cannot possibly justify the independent invention hypothesis. In all those cases where both hypotheses are possible, there seems to be an almost automatic preference for independent invention. The reasons for this are by no means clear.

The discovery of a good counting system, which enables people to count better and more easily than their forebears did, means a great advance in civilisation. Such discoveries are very rare: there are still tribes who cannot count beyond 5 or 6. Also, better counting systems have a tendency to spread and to replace more primitive methods.

For these reasons, therefore, there is a preference for the diffusion hypothesis here. If two counting systems are found to be very similar in many details, the diffusion hypothesis immediately explains why they are similar, whereas the hypothesis of independent invention supplies no direct explanation at all. The main purpose is to present the relevant facts. Certain hypotheses (chiefly those of Seidenberg) are explained and the supporting arguments discussed but it is important to remember that other hypotheses, in particular that of independent invention, are also possible.

Written numbers

Having discussed *counting systems* and *number words* in Chapters 2 and 3, *written numbers* (or *numerals*) are examined in Chapter 4. The primary concern in this latter chapter is with systems for writing down *whole numbers*; *fractions* and *methods of calculation* are considered in Chapters 5 and 6. This approach will enable the reader to make comparisons between different cultures, and to understand the principles involved all the more easily. Cross-references will enable the reader to look at several aspects of one culture and to see how the different cultures responded to a common problem.

Chapter 4 begins by discussing numeral systems of civilisations of the past. Later the history of the familiar Hindu-Arabic numerals is traced, from their origin in the Indian subcontinent to their gradual adoption in the West.

Fractions and calculation methods

In Chapter 5, both *fractions* and *calculation methods* are discussed: these two topics cannot be separated, as will become evident. Four quite different methods of writing fractions are known. Each of the four methods is associated with a particular way of performing calculations. The characteristic features of the four systems are as follows.

1. The ancient *Egyptians* had a systematic notation for *unit fractions* 1/n only. Mixed fractions *m/n* were expressed as sums of unit fractions. There are a few exceptions: they had special notations for 'natural fractions' (like 2/3 and 3/4) which frequently occur in everyday life.
2. The *Babylonians* systematically used sexagesimal fractions, that is fractions with denominator 60 or 60^2 or 60^3, etc. They performed all calculations in the sexagesimal system.
3. *Mixed fractions m/n* were used by the Greeks, Chinese, Hindus and Arabs. We all have learnt to use these fractions at school, but our engineers and scientists seldom use them: they prefer decimal fractions.
4. *Decimal fractions* came into general use in Western Europe at the end of the sixteenth century. Their early history in China and elsewhere will be discussed briefly.

There is no point in explaining the Egyptian method of writing fractions without explaining also how the Egyptians managed to perform additions, multiplications and divisions, using unit fractions only. For instance: How did they multiply 1/5 by 4, if they could not write the result as 4/5?

When dealing with the Babylonian system, the same situation arises. The Babylonian notation for sexagesimal fractions can be explained in a few words, but after this one has to consider how calculations were made in this system.

The discussion therefore in Chapter 5 includes *Egyptian fractions* and methods of calculation, *Babylonian sexagesimal fractions* and methods of calculation, *Greek, Chinese and Hindu mixed fractions* and methods of calculation, and, finally, *decimal fractions*.

Aids to calculation

The final chapter covers the history of the *abacus* (or counting board), early *calculating machines*, and, lastly, the evolution of the modern *electronic computer*.

Number, as already stated, plays an essential role in our culture. This book is intended to stimulate a deeper understanding both of number itself and of the evolution of certain fundamental ideas associated with the communication of numerical information. It is envisaged therefore primarily as a resource book of ideas for those who are fascinated by numbers and everything associated with them, and equally (or perhaps especially) for those who have the task of communicating numerical ideas to students both young and not so young.

2 Counting Systems

Decimal counting

Our counting system is based on the number ten. Why just ten? Probably because we have ten fingers. Finger counting is widespread among both primitive and civilised peoples (see Figure 1).

The main principle of the decimal system is that ten is considered as a new unit, from which point counting starts again. The multiples of ten are counted by the same system: ten, twenty, thirty, etc. Ten tens are again a new unit and so on.

This way of counting is very old. English is a Germanic language and the oldest known Germanic language, namely Gothic, the language of the Bible translation of the Bishop Wulfila (c. AD 350), already has the same method of counting. Essentially the same system is also found in all Romance languages, such as French and Italian, which are derived from Latin. The Greek language, which existed as early as 1200 BC, has the same counting method and so has Hittite, which was being

FIGURE 1   Finger numbers from a thirteenth-century Spanish codex, in Luca Paccioli's *Summa de arithmetica, geometria proprotioni et proportionalita*, 1494 (British Museum)

spoken in Asia Minor by 1800 BC. All these languages belong to the 'Indo-European' family of languages. As 'Indo-European' indicates, these languages were (and still are) spoken in a region ranging from India to Europe. They are supposed to stem from one mother language, the 'original Indo-European language'. This original language must have come into existence between 3000 and 2500 BC. Since all Indo-European languages have the same counting system and very similar number words (compare e.g. 'two' and 'three' with Latin *duo* and *tres*), one may safely conclude that the Indo-European mother language had the same method of counting.

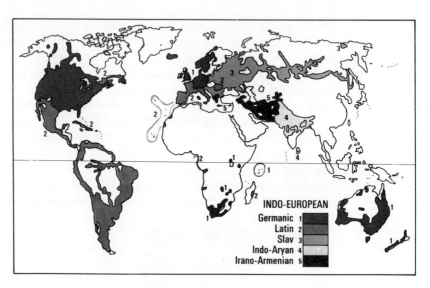

FIGURE 2   Distribution of Indo-European languages today

This method is called the *decimal counting system*. It should be clearly distinguished from the *decimal number notation*. Spoken words like 'twenty-three' or 'three-and-twenty' belong to the decimal counting system, whereas written numerals like '23' belong to the decimal number notation, which is comparatively recent, as will be seen later. For now we are concerned with counting systems only and not with systems of numerals. Numerals will be considered later.

The origins of our decimal counting system are hidden in obscurity because Indo-European languages were spoken long before they were written. What can be observed up to the present is the spread or diffusion of this counting system from more civilised to less civilised people. Whoever wants to trade with the United Kingdom has to learn its money system, which is in turn closely connected with the method of counting. The black population of North America speaks English and uses the decimal system, whereas their ancestors in Africa had more primitive number systems. Quite generally, more efficient counting systems tend to spread and more primitive methods tend to disappear.

Different counting methods, such as those used by ancient civilised peoples and by primitive tribes, will now be compared with the decimal counting system to see if there are special ways in which they are related.

Primitive counting methods

We start with the Bergdama tribe in South Africa. According to the ethnologist H. Vedder[1], most Bergdama count by independent number words up to 10 but these were taken over from the Nama language. The older people used words that Vedder could not make out, but one Bergdama woman told him that the only number words which she knew were 'one', 'two' and 'many'. She also said that her people in the mountains knew no other number words.

This example shows how, in the course of one or two generations, a very primitive and restricted counting method was replaced by a better system, which the Bergdama learned from their neighbours, the Nama.

Seidenberg, in 'The Diffusion of Counting Practices', states:

Counting may seem to be an elementary process that any human being would come across by himself, yet the fact that many groups of savages cannot count beyond 2 makes us realize that the process is an advanced one. Over most of aboriginal Australia one finds essentially only two number words, 'one' and 'two': many of the tribes are reported definitely as not counting beyond 2, and as indicating higher multiplicities by the word 'many', while others go a little further by compounding these words – for example, expressing 3 as 'two-one', 4 as 'two-two', and 5 as 'two-two-one'. A similar method of counting, the so-called 2-system, is found in New Guinea, in South America, and in South Africa.[2]

The fact that some Australian aboriginal tribes, for example, cannot count beyond 2 is indicative of a comparatively primitive state of civilisation. Contact with a more advanced state soon demands that such a limited counting system be extended. These peoples may have been able to count on their fingers without using words, but as soon as it becomes necessary to trade with goods or money the need for more number words becomes imperative.

A somewhat higher level of counting is reported of the Kalahari Bushmen in Africa. S. S. Dorman writes:

They have no words beyond five, and not often beyond three, for they cannot count any farther. I have satisfied myself on that point by putting some Kalahari Bushmen to the test on more than one occasion.[3]

Counting up to 3 or 5 with no facility for any further detailed counting cannot be considered to be a counting system. A set of independent number words like 'one, two, three' or 'one, two, three, four, five', is known as *systemless counting*. On the other hand, a method which forms composite number words like 'fifteen' or 'fifty', is referred to as a 'counting system'.

The following is an extract from a paper on the tribesmen of central Brazil:

We turn to the primitive Bakairi tribe with whom I made different arithmetic experiments.

The names of their calculation 'equipment', their fingers, have nothing to do with numbers. The thumb is called 'father', and the little finger is called 'child' or 'little one'. The middle finger is, as with us, the 'middle'; indexfinger and ringfinger become 'neighbours' – that is, at least, the obvious interpretation – of the 'father' and the 'little one'.

They count in words up to 6:

1 = *tokále*	4 = *aháge aháge*
2 = *aháge*	5 = *aháge aháge tokále*
3 = *aháge tokále* or *ahewáo*	6 = *aháge aháge aháge*

It is noticeable that the numbers 1 to 3 have their own names, and that the names for the numbers 4 to 6 are composites of *aháge* and *tokále*. I became acquainted with the word *ahewáo* for 3 on my second expedition, but it is used as often as the composite 'two-one'. The idea that *ahewáo* might be a new word is wrong; comparing the languages proves that the older Karaibi tribes, north of the Amazon, had previously used *ahewáo*. It is noticeable that 3 (*ahewáo*) is used in none of the higher figures, not even in the number word for 6.

When we remove the independent word *ahewáo*, the Baikairi counts:

one	two-two
two	two-two-one
two-one	two-two-two

Beginning with the little finger of the left hand he says *tokále*, then he joins fingers IV and V and says *aháge*. He keeps finger III separated form IV and V and says *aháge tokále*; he joins fingers II and III and says *aháge aháge*, touches the thumb and says *aháge aháge tokále*, and pushes the little finger of the right hand against the fingers of the left hand and says *aháge aháge aháge*. After 6 the Baikairi is not able to count further; he just touches fingers IV, III, II and I of the right hand and repeats *méra* ('this one'), then he touches the toes of his left foot and after that the toes of his right foot, and with every touch he says *méra*. If he wants to express more than 20, he tousles his hair in all directions saying *méra méra*.

Logically, they should call 7 *aháge aháge aháge tokále*, but they would have to count the number words themselves and they cannot do this. Even when it affects important matters, their lack of words and their inability even to indicate higher amounts than 20 is an invincible obstacle. Obviously after 6 they have only a vague approach to higher amounts. Imagine that somebody is climbing a staircase and for some reason intends to count the steps, but suddenly realizes that he does not know for certain how many steps he has climbed so far. Disconcerted, he gets more and more mixed up and so has no idea how many steps he has climbed. The Baikairi climbs the staircase knowing that after the toes of his first foot, he climbed more steps than the fingers of his left hand, after the toes of the second foot he climbed more steps than after the toes of his first foot. That 10 fingers and 5 toes represent 15 is beyond their imagination.

Their conception after 6 is only clear to the extent that they can touch fingers and toes. As soon as precise amounts have to be counted their ruses hardly reach further than their knowledge of figures. I have tried experiments in the Paleko and Tumayaua districts—I could tousle my own hair when I think about it—but will mention just the example of a large amount of corn seeds. When I put ten corn seeds in front of them and asked *atúra* ('how many'), they counted them in the way described above, slowly but correct up to 6. For the seventh and eighth corn seed they showed fingers IV and V of the right hand and said *méra méra*, but after that they were exhausted. They reminded me of people who are playing cards without interest and soon yawn and say 'I have no talent for playing cards'. The Baikairis yawned as well and when I urged them to carry on they pulled faces, moaned about *kinaráchu zwáno* which means 'brain work' but can also mean 'headache', and sometimes they just walked out. In any event, they went on strike.[4]

From this evidence we see that the Bakairis are effectively obliged to pair off objects to be counted. When asked to count or give the number of as few as three grains of corn, it is recorded that they separate them into a pile of two and the remaining one. In one of their folk-tales, in order to express the fact that the hero felled five trees, they use three sentences: 'He felled two trees. Again he felled two trees. He felled another.'

Call this method *counting by pairs*. It is closely connected with a set of number words formed by repeating the words for 1 and 2. There was another example in the extract above from Seidenberg's paper, quoted earlier in this chapter. This system of counting will be discussed by looking at three examples taken from the same paper (see Table 1).

TABLE 1   Counting by pairs

	Gumulgal (Australia)	Bakairi (South America)	Bushman (South Africa)
1	*urapon*	*tokale*	*xa*
2	*ukasar*	*ahage*	*t'oa*
3	*ukasar-urapon*	*ahage tokale* (or *ahewao*)	*'quo*
4	*ukasar-ukasar*	*ahage ahage*	*t'oa-t'oa*
5	*ukasar-ukasar-urapon*	*ahage ahage tokale*	*t'oa-t'oa-ta*
6	*ukasar-ukasar-ukasar*	*ahage ahage ahage*	*t'oa-t'oa-t-oa*

In Africa, *pure 2-count* is found only among the Bushmen. They count by this method up to $10 = 2 + 2 + 2 + 2 + 2$. Beyond 10 the method does not work: the number words become too long. To see this, consider how a listener, having only a pure 2-count, would interpret 'two and two and two and two and two and two and two and two'. He would need a more sophisticated counting system in order to count how often the word 'two' has been said.

In Australia, pure 2-count is used by several tribes in the east, and it is also found in nearby New Guinea (see Figure 3, black circles). This region is small and connected and all tribes in it speak closely related languages. It is hardly a reasonable hypothesis that pure 2-count was independently invented five times in one region having the same language group throughout, so one may reasonably suppose that the method was invented only once and then spread over the eastern region of Australia and New Guinea.

Among the South American Indians pure 2-count is found in a connected region in the north-east. Here too it seems reasonable to suggest that it was invented only once in South America. Seidenberg explains the striking uniformity in the formation of number words in Africa, Australia and South America by making an even bolder guess:

> Another possible explanation of the uniformity is that the 2-system had a single origin. From this point of origin it spread out over the whole earth; later, other methods of counting arose and spread out over almost all, but not quite all, of the world. Note that the 2-system appears now only at the edges and seems ready to be wiped off the face of the globe.[5]

This hypothesis is very interesting, but the available evidence is not sufficient to prove it, as Seidenberg himself admits. On the other hand, Seidenberg is certainly right in assuming that in ancient times 2-count was in use in a much larger area than it is now. Of course, the evidence as presented so far is also not inconsistent with the *independent invention* hypothesis. The *diffusion* hypothesis with its concept of a

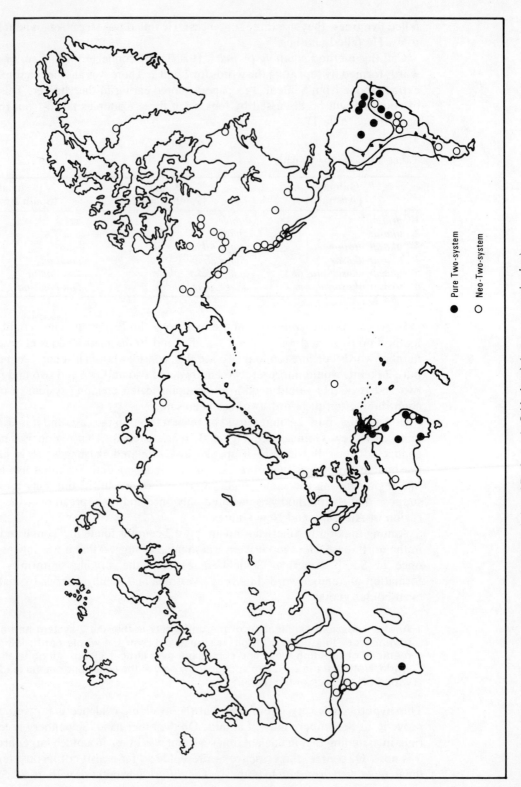

FIGURE 3   Pure 2- and neo-2-counting (by kind permission of Professor A. Seidenberg) (see accompanying key)

Key to Figure 3

Pure 2-Counting
(with at least 2+2, 2+2+1)

South America
Left to right:

Zamuco	2+1, 2+2, ..., 2+2+2+1
Arara	2+1, 2+2, 2+2+1
Bakairi	2+1, 2+2, 2+2+1, 2+2+2
Cayapo	2+1, 2+2, 2+2+1, 2+2+2
Crengez	2+2, 2+2+1
Apinage	2+1, 2+2, 2+2+1
Krao	2+1, 2+2, 2+2+1, 2+2+2
Canella	2+1, 2+2, 2+2+1, 2+2+2
Cotoxo	2+1, 2+2, 2+2+1

Australia
Bottom to top:

South Narrinyeri	2+2, 2+2+1, 2+2+2, 2+2+2+1, etc.
North Kuri	2+1, 2+2, 2+2+1
Minyung	2+1, 2+2, 2+2+1, 2+2+2, etc.
Kana	2+1, 2+2, 2+2+1
(Gualluma	2+2, 2+2+1; see Neo-2-Counting)
Kohoyimidir	2+1, 2+2, 2+2+1
Kauralgal	2+1, 2+2, 2+2+1, 2+2+2, 2+2+2+1, 2+2+2+2
Gumulgal	2+1, 2+2, 2+2+1, 2+2+2
Torres Straits	2+1, 2+2, 2+2+1, 2+2+2

New Guinea

Parb	2+2, 2+2+1
Sisiami	2+1, 2+2, 2+2+1
Anal; Arop	2+1, 2+2, 2+2+1

South Africa

Bushman	2+2, 2+2+1, ..., 2+2+2+2

Neo-2-Counting

South America
Bottom to top:

Tsonega	$2 \times 3, 2 \times 4, 2 \times 5$
Tehuelche	$2 \times 3, 2 \times 4$
Guarani	$3 \times 2, 4 \times 2, 2 \times 6, 6+7, 2 \times 9, 2 \times 9+1, 4 \times 5, 2 \times 4 \times 5$
Toba	$2+1, 2+3, 2 \times 3, 1+2 \times 3, 2 \times 4, 2 \times 4+1$
Caingua	$3 \times 2+1, 2 \times 4$
Payagua	$2 \times 3, 2 \times 4, 2 \times 4+1, 2 \times 5$
Cherente	$3 \times 2, 4 \times 2$

Note: The fainter, but still fairly clear, trace 2×3 is found in the region between the pure 10-counters and 5–20-counters amongst the following:

Tiatinagua	2×3
Zapara	$2 \times 2, 2 \times 2+1, 2 \times 3, 2 \times 3+1$ (to 7 only)

Also with the:

Ona	$2 \times 2, 2 \times 3$
Marawan	2×3 (6 contains 3)

North America
West Coast, bottom to top:

Yuma	$3 \times 2, 4 \times 2$
Paiute	$3 \times 2, 4 \times 2$
Tobikhan	$3 \times 2, 2 \times 4, 2 \times 5$
Netela	$3 \times 2, 4 \times 2$
Salinas	$3 \times 2, 8-1, 4 \times 2, 10-1$
Wintun	$2 \times 3, 6+1, 2 \times 4, 8+1, 2 \times 5$
Yana	$3 \times 2, 4 \times 2$

Tehama	$2 \times 3, 2 \times 4$
Haida	$2 \times 2, 3 \times 2, 4 \times 2$
Kygani	$2 \times 2, 3 \times 2, 4 \times 2$

Interior, left to right:

Tukudh	$2+1, 2 \times 3, 2 \times 4$
Kutchin	$2 \times 3, 2 \times 4$
Tacully	$2 \times 3, 2 \times 4$
Dogrib	$2 \times 3, 2 \times 4$
Sarcis	$2 \times 3, 2 \times 4$
Sicanni	$2 \times 3, 2 \times 4$
Beaver	$2 \times 3, 2 \times 4$
Montagnais	$2 \times 3, 2 \times 4$
Chippewa	$2 \times 3, 2 \times 4$
Digger	
(Huerfano Park)	$2 \times 3, 2 \times 4$
Atakapa	
(on the Gulf)	$3 \times 2, 4 \times 2$

Eskimo:

Labrador	$3 \times 2, 4 \times 2, 5 \times 2$
Melville Bay, Greenland	$3 \times 2, 3+4, 4 \times 2, 4+5, 5 \times 2$

Asia

Yukaghir	$2 \times 3, 2 \times 4$
Tsuihoan (Formosa)	$2 \times 3, 2 \times 4$
Karen (South Burma)	$3 \times 2, 3 \times 2+1, 4 \times 2+1$

Nicobar Islands:

Inland tribes	3×2
Coast tribes	2×4

Pacific

Radak Group, Marshall Islands	$2 \times 3, 2 \times 3+1, 2 \times 4+1$

New Guinea:

Waima, Roro, Kabadi, Pokau	$2 \times 3, 2 \times 3+1, 2 \times 4, 2 \times 4-1$
Hula, Keapara, Galoma	$2 \times 3, 2 \times 4-1, 2 \times 4, 10-1$

Australia, right to left:

Kamilaroi	$2+2, 2+3, 3+3$
Wirri-Wirri	$2+1, 2+2, 3+2, 3+3$
Gualluma (Upper Sherlock River)	$2+2, 2+2+1, 3+3$

Africa
Sudan, left to right:

Gurunsi	$6+1, 8+1$
Munsi	$3+3, 4+4$
Ekio, Akparabon, Ododop, *et alia*	$3+3, 4+3, 4+4, 5+4$
Sao Group (Kotoko of Kousri, of Logone, of Goulfei; Logone)	$2 \times 3, 2 \times 4$
Mbai, Sara Gulei	$3+3, 4+3, 4+4, 4+5$

Boa Group:

Niellim	$2 \times 3, 2 \times 4$
Tounia	$3 \times 2, 4+4$
Ndonga	$2 \times 3, 2 \times 4$
Mongwandi	$3+3, 4+3$
Kanderma	$2 \times 3, 4+3$

Bantu, top to bottom:

Bali (Cameroons)	$6+1, 8+1$
Kombe	$6+1, 8+1$
Luba-Hemba	$8-1, 10-1$
Konde	7, 8 bracketed; 9. 10 bracketed

single centre of invention demands that plausible evidence be produced as to where that centre might be. But are the two hypotheses necessarily mutually inconsistent? Seidenberg considers these points.

We are confronted with two types of explanation, the 'diffusionist' and the 'independent inventionist'. It may be asked whether the two types are not, after all, quite consistent; and whether perhaps the 2-system, having been generated in one of the ways suggested by the 'independent inventionists', then spread over the earth from a single center, as required by the 'diffusionists'. Here I need to make clear that if I insist upon diffusion from a *single* center it is not because I am especially fond of the number 1, but because if the diffusionist view is correct, then we may look for the origin of the 2-system in that center.

And where will that center be? The 2-system appears to embody an idea; it is a form of knowledge. Knowledge always passes from those who know to those who do not know, not the other way. This implies that knowledge will usually pass from higher centers of culture to lower ones, though the direction may sometimes be from lower to higher, in analogy with the fact that water seeks its level; in any event, the starting point will be a center of knowledge. It is known that ideas arose in the areas of the ancient civilizations; it is not known that ideas ever arose, anciently, anywhere else. Therefore, we will look in the ancient centers of civilization for the origin of the 2-system.

. . . .

The theory of a single origin of counting explains the uniformity without explicitly explaining the 2; the theory of independent invention, on the other hand, is obliged immediately to ponder the 2, and soon comes to the suggestion that we psychoanalyze ourselves to find out what the savage really knows. The diffusionist theory, for its part, having indicated historical reasons for the uniformity, is committed to a historical research.[6]

Note: The 'ways suggested by the independent inventionists' (line 4 of the above extract) include the 'duality' concept (by which we distinguish another person from ourselves), the 'two-sided' concept (by which we distinguish what we *can* see 'in front of us' from what we *cannot* see 'behind us'), and the 'pair-wise' concept (by which we recognise in eyes, hands, feet, etc., the idea of 'left and right').

In support of his hypothesis Seidenberg adduces evidence from ancient Egyptian, Sumerian and Persian documents. We shall now present the Persian evidence; the Sumerian and Egyptian number notations will be discussed later.

In the Behistûn inscription of Darius the Great (*c.* 500 BC) there are the following number signs.

The old Persian number notation is quite similar to the Babylonian method of writing numbers up to 10 as shown below. (See also Figure 4.)

The signs for 1 and 10 are the same in both systems and the numbers from 2 to 9 are written by repeating the simple 'wedge' for 1. However, in Babylonia, the 'wedges' were placed in horizontal groups of 3, whereas in the Persian system they were arranged in vertical groups of 2. The scribes who wrote the Persian text of the inscription were familiar with Babylonian cuneiform script and they adapted it to the Old Persian language. Now the question arises: Why did they change the form of the number signs for 7, 8 and 9, arranging the wedges in pairs?

The Babylonian sign for 9 can be recognised at a glance whereas in reading the Persian signs for 8 and 9 one has to count the wedges in the first or second row.

FIGURE 4   Part of Babylonian clay tablet
(British Museum)

Why did the Persian scribes replace the perfect Babylonian sign for 9 by a less practical arrangement of the wedges in pairs? The most plausible hypothesis would seem to be that they adapted the Babylonian number notation to their own counting habit, *counting by pairs*. This is the hypothesis which Seidenberg[*] puts forward. It is, of course, true that the Persian number words of the time were purely decimal and show no trace of a 2-system. Nevertheless, their earliest counting habit may well have been counting by pairs.

In pure 2-count, 6 and 8 are pronounced as

$$6 = 2 + 2 + 2 \text{ and } 8 = 2 + 2 + 2 + 2$$

This is very inconvenient. Another method, which is similar but more efficient, uses special words for 3 and 4 and pronounces 6 and 8 as 'twice three' and 'twice four'. Thus, the Toba in Paraguay have the following number words[7]:

1	*nathedac*
2	*cacayni* or *nivoca*
3	*cacaynilia*

4   *nalotapegat*
5 = 2 + 3   *nivoca cacaynilia*
6 = 2 × 3   *cacayni cacaynilia*
7 = 1 + (2 × 3) *nathedac cacayni cacaynilia*
8 = 2 × 4   *nivoca nalotapegat*
9 = (2 × 4) + 1 *nivoca nalotapegat nathedac*
10 = 2 + (2 × 4) *cacayni nivoca nalotapegat*

Seidenberg calls this method *neo-2-counting*. It has several variants, which may well be independent inventions. All these variants have in common the decompositions

$$6 = 2 \times 3 \text{ (or } 3 + 3)$$
$$8 = 2 \times 4 \text{ (or } 4 + 4)$$

but 7 and 9 are sometimes expressed as

$$7 = 6 + 1 \text{ and } 9 = 8 + 1$$

and in other cases as

$$7 = 8 - 1 \text{ and } 9 = 10 - 1$$

or even as

$$7 = 4 + 3 \text{ and } 9 = 5 + 4$$

On the map in Figure 3 neo-2-counting is indicated by circles, with no distinction between variants being made. This shows that neo-2-count is widespread in Africa, southern Asia, Australia and the Americas. In South America the expressions 7 = 6 + 1 and 9 = 8 + 1 prevail (see the example of the Toba quoted above) whereas in Africa divisions into nearly equal parts, such as

$$7 = 4 + 3 \text{ and } 9 = 5 + 4$$

are quite frequent. It can also be seen that in Australia and New Guinea as well as in South America the regions of neo-2-count are adjacent to areas of pure 2-count. To explain this, Seidenberg assumes that all these regions originally had pure 2-count, and that later on this very inconvenient counting method was replaced by the more efficient methods of neo-2-count.

Can one explain the development of neo-2-count from pure 2-count? It might seem that the decompositions 6 = 3 + 3 and 8 = 4 + 4 are totally different from 6 = 2 + 2 + 2 and 8 = 2 + 2 + 2 + 2. Still one can imagine a passage from pure-2-count to neo-2-count through the intermediate of tally sticks or written numbers. Suppose that some pure 2-counting people would memorise and transmit numbers by carving notches in a tally stick. Since they counted by pairs, their carved numerals would probably look like this:

$$6 \; \begin{matrix} |\,|\,| \\ |\,|\,| \end{matrix} \qquad 7 \; \begin{matrix} |\,|\,|\,| \\ |\,|\,| \end{matrix} \qquad 8 \; \begin{matrix} |\,|\,|\,| \\ |\,|\,|\,| \end{matrix}$$

This is not pure fancy for numerals of just this kind were actually written by the Egyptians as early as 3000 BC, as will be shown in Chapter 4. The earliest Sumerian numerals, written on clay, were of the same type.

Now suppose that carved or written numerals of this kind were used in ancient times by a 2-counting population, and that this method of noting numbers was taken over by neighbours who had independent number words for 1, 2, 3 and 4. They might read the sign for 6 as 3 + 3, the sign for 7 as 4 + 3 and the sign for 8 as 4 + 4. In this way the passage from pure 2-count to neo-2-count might be explained.

Seidenberg also discusses the Sumerian and Egyptian methods of writing numbers, which were in use about 3000 BC. He regards them as giving evidence for the existence of 2-count or neo-2-count in the Near East about 3000 BC. The evidence is not conclusive, but consider the facts. Figure 5 shows a reproduction of a Sumerian clay tablet dating from about 3000 BC or earlier. In the middle column of the tablet are (from bottom to top) the numbers

$$1, 2, 3, 4, 5, -, 7, 8, 9 \text{ (6 is broken off)}$$

From 4 upwards the numbers are divided into equal or nearly-equal parts, as in a neo-2-system:

$$4 = 2 + 2, 5 = 3 + 2, 7 = 4 + 3, 8 = 4 + 4, 9 = 5 + 4$$

Essentially the same method of writing numbers is found in Ancient Egyptian texts. Here too the numbers below 10 are formed by repeating the sign for 1 – a vertical stroke – as often as necessary as shown below.

Seidenberg's hypothesis is that this variant of the neo-2-system spread out from the ancient civilisations of Mesopotamia and Egypt towards the middle zone of Africa, where it is still found today.

The two counting methods just explained, 2-count and neo-2-count, do not carry us very far; beyond 10 they become impracticable.

A slightly better system is *4-count* – counting in groups of four. This system is not very common. Seidenberg gives two examples from Africa. The number words of the Afudu and Huku are equivalent to:

Afudu	Huku
$6 = 4 + 2$	$8 = 2 \times 4$
$7 = 4 + 3$	$9 = 2 \times 4 + 1$
$8 = 4 \times 2$ or $4 + 4$	$13 = 12 + 1$
$9 = 4 + 5$	$14 = 12 + 2$
	$15 = 12 + 3$
	$16 = (2 \times 4) \times 2$
	$17 = (2 \times 4) \times 2 + 1$
	$18 = (2 \times 4) \times 2 + 2$
	$19 = (2 \times 4) \times 2 + 3$

In South America, the Lulu count by the 4-system at least to 9 and the Charrua and Mocovi at least to 8.

From the number words of the Huku it can be seen that in their counting the number 12 is used as a stepping-stone from which the counting starts anew. In other

FIGURE 5   Sumerian clay tablet
(Staatliche Museen zu Berlin,
DDR) and map showing the
location of Sumer

languages also the number 12 plays a special role. We often count by the *dozen*; the word is derived from the French word *douzaine*, which was actually an alternative term for the French shilling, the *sou*. In France, the old table of monetary equivalents was:

$$1 \; livre = 20 \; sou$$
$$1 \; sou \;\; = 12 \; denier$$

(This was derived from Charlemagne's monetary standard of 780, which embodied a basic 12-unit:

$$1 \text{ } \textit{libra} \text{ or } \textit{talentum} = 20 \text{ } \textit{solidus}$$
$$1 \text{ } \textit{solidus} \qquad = 12 \text{ } \textit{denarius})$$

The *sou*, being equivalent to 12 *denier*, thus came also to be called a *douzaine* (a 'twelver'). We call a dozen dozens a *gross*; this word is derived from the French *grosse douzaine* (a 'strong dozen'). The dozen and the gross are both still used for counting, chiefly in connection with commercial transactions. It is also known, for example, that the Romans divided their main weight, the *as*, into 12 *unciae* (or ounces) and their measure of length, the *pes* (foot), into 12 *pollices* (inches). The most obvious reason for the special role of the number 12 in connection with weights and measures is its exact divisibility into half, quarter and third.

In addition to its divisibility, there is another possible reason for the special role of 12 as a measure of quantity. This relates to the idea of 'excess' of which there are examples in the *1001 Arabian Nights* and the French use of *quinze jours* for 'fortnight'. Although these two examples illustrate an excess of 1 only, it is possible to produce evidence suggesting the existence of cases of an excess of 2. Thus, the special role of 12 cannot be used to defeat the argument that *counting by tens is 'natural' because of the ten fingers on a person's hands.*

There is also much evidence to suggest that the number 12 has often had mystical or special symbolic significance. There are the twelve Christian apostles, the twelve ships of each of the heroes Ajax and Odysseus in Homer's *Iliad* and so on.

In the decimal system numbers are expressed as sums of powers of ten with coefficients ranging from zero to nine. Thus 209 can be written as

$$2 \times 10^2 + 0 \times 10^1 + 9 \times 10^0 \text{ (Note that } 10^0 = 1)$$

Instead of ten, any other positive integer (e.g. 2 or 5) might have been used as a base of the system. Of special interest, because it is used in electronic digital computers, is the *binary* or *dyadic system* which has 2 as its base.

In this system, numbers are expressed as sums of powers of 2 (including $2^0 = 1$) with coefficients 0 or 1, for instance

$$13 = 8 + 4 + 1 = 1 \times 2^3 + 1 \times 2^2 + 0 \times 2^1 + 1 \times 2^0$$

The coefficients of the successive powers of 2 are 1, 1, 0, 1. Hence the number 'thirteen' is written in the binary system as 1101.

The procedure to express any given number N as a sum of powers of 2 is as follows. Let 2^a be the highest power of 2 less than or equal to N. Subtract 2^a from N. The difference $N - 2^a$ is less than 2^a. If the difference is zero, N is just 2^a. If not, repeat the procedure, again subtracting the largest possible power 2^b, and so on until the difference becomes zero. At the end

$$N = 2^a + 2^b + \ldots + 2^i$$

For instance, if N is 22, then

$$22 = 16 + 4 + 2$$
$$= 1 \times 2^4 + 0 \times 2^3 + 1 \times 2^2 + 1 \times 2^1 + 0 \times 2^0$$

hence the binary notation for 'twenty-two' is 10110.

There is no evidence to suggest that the binary system was used to form number words. It is easy to find a reason for this: the base 2 is too small for everyday practical use. The decomposition of 30, for example, into $16 + 8 + 4 + 2$ is too cumbersome. However, computers use the binary system for arithmetic calculations.

The 'memory' of a computer can retain only successions of zeros and ones, nothing else. If the button marked 5 on a computer is pressed, it immediately transforms the number 5 into the sequence 1–0–1 according to a transformation rule fixed once and for all:

> Press button marked 1, the computer notes 1
> press button marked 2, the computer notes 1–0
> press button marked 3, the computer notes 1–1

and so on.

Now what happens if a higher number, say 240, is fed into the computer? First press the button marked 2; the computer notes the binary equivalent 1–0. Next press the button marked 4. The computer now automatically multiplies the number 1–0 (= two) just noted by 1–0–1–0 (= ten) which gives 1–0–1–0–0 (i.e. twenty) and adds 1–0–0 (= four). The result of this addition in the binary system is 1–1–0–0–0 (= $2^4 + 2^3 = 24$), which is correct because at this stage 24 has been fed into the computer. Finally press the button marked 0. The computer again multiplies the result of the preceding operations, i.e. 1–1–0–0–0, by 1–0–1–0 (= ten because ten is the base of our system) and it adds 0. Thus, the computer has in binary notation

$$1\text{–}1\text{–}0\text{–}0\text{–}0 \times 1\text{–}0\text{–}1\text{–}0 = 1\text{–}1\text{–}1\text{–}1\text{–}0\text{–}0\text{–}0\text{–}0$$

which means in decimal notation

$$24 \times 10 = 240$$

The binary product is:

$$2^7 + 2^6 + 2^5 + 2^4 + 0 + 0 + 0 + 0 = 128 + 64 + 32 + 16 = 240$$

One must distinguish between *pure 2-count* and the *binary system* of numbers. *Pure 2-count* is a system of counting having numbers words for 1 and 2 only. Words for numbers greater than two are formed compositely. Thus 3 is called 'two-one', 4 is called 'two-two', and so on. The *binary system* of numbers, on the other hand, is a place-value system of recording numbers used in computers and similar to our own decimal place-value system, but having the base 2 instead of the base 10. There is no evidence that a binary counting system has ever existed.

In many parts of the world, counting by independent number words stops at 5. Between 5 and 10, composite number words are used. Thus, for example, the Aztecs of Mexico count as follows:

1 *ce,*	hence 6 *chica-ce* $(6 = 5 + 1)$
2 *ome,*	hence 7 *chic-ome* $(7 = 5 + 2)$
3 *yey,*	hence 8 *chicu-ey* $(8 = 5 + 3)$
4 *naui,*	hence 9 *chic-naui* $(9 = 5 + 4)$
5 *macuilli*	

In the composite number words for 6, 7, 8 and 9 the first part *chica* obviously means either 'five' or 'one hand' or something like this, although *chica* cannot be derived from *macuilli* = 5.

This method of counting, which is widespread among the indigenous peoples of the Americas, Asia and Africa, may be called *5-count*. It can easily be extended to numbers between 10 and 20. Thus the Aztecs continue their count as follows:

10 *matlactli*
11 *matlactli-on-ce* (11 = 10 + 1)
12 *matlactli-on-ome* (12 = 10 + 2)
13 *matlactli-on-yey* (13 = 10 + 3)
14 *matlactli-on-naui* (14 = 10 + 4)
15 *caxtulli*
16 *caxtulli-on-ce* (16 = 15 + 1)
etc. until
19 *caxtulli-on-naui* (19 = 15 + 4)

Beyond 20 there are two methods of extending the system, both of which occur frequently. Following Seidenberg, these methods can be called the *5–10-system* and the *5–20-system*.

In the 5–10-system multiples of ten are constructed simply as multiples of ten not as multiples of twenty. For instance in the 5–10-system 80 is just 8 times ten and not *quatre-vingt* and 90 is 9 times ten and not *quatre-vingt-dix*. The 5–10-system is mainly found in Africa and North America (see Figure 13) and is very often connected with finger-counting. In many cases the number words are derived from the names of the fingers.

In the 5–20-system numbers like 40 and 60 are pronounced as 2×20 and 3×20. The Aztecs provide a typical example of a fully developed 5–20-system. Their number-words for 20, 30, etc. are as follows:

20 *cem-poualli* (20 = 1×20)
30 *cem-poualli-om-matlactli* (30 = $1 \times 20 + 10$)
40 *ome-poualli* (40 = 2×20)
100 *macuil-poualli* (100 = 5×20)

There are also special words for powers of 20:

$$20^2 = 400 \quad cen\text{-}tzuntli$$
$$20^3 = 8000 \quad cen\text{-}xiquipilli$$

A closely related system is that of the Mayas of Yucatan. Their number words up to 19 are as follows:

1 *hun*	9 *bolon*
2 *ca*	10 *lahun*
3 *ox*	11 *buluc*
4 *can*	12 *lah-ca* (10 + 2)
5 *ho*	13 *ox-lahun* (3 + 10)
6 *uac*	14 *can-lahun* (4 + 10)
7 *uuc*	15 *ho-lahun* (5 + 10)
8 *uaxac*	. . . . .
	19 *bolon-lahun* (9 + 10)

The Mayas have special words for powers of 20:

$$(1 \quad hun)$$
$$20 \quad hun\text{-}kal$$
$$20^2 \quad hun\text{-}bak$$
$$20^3 \quad hun\text{-}pic$$
$$20^4 \quad calab$$

The words for multiples of 10 below 400 are as follows:

30 *lahu-cakal* ($2 \times 20 - 10$)   100 *ho-kal* (5×20)
40 *ca-ikal* (2×20)   120 *uac-kal* (6×20)
50 *lahu-y-oxkal* ($3 \times 20 - 10$)   140 *uuc-kal* (7×20)
60 *oxkal* (3×20)   200 *lahun-kal* (10×20)
70 *lahu-cankal* ($4 \times 20 - 10$)   300 *ho-lhu-kal* (15×20)
80 *can-kal* (4×20)
90 *lahu-y-hokal* ($5 \times 20 - 10$)

FIGURE 6   Mayan day and month signs (British Museum)

The Mayan number sequence is unusual and almost certainly did not arise from the everyday needs of the common people. Menninger[8] suggests that it was an artifical creation of the priests, possibly intended for calendar computations.

Unlike the Aztecs the Mayas have independent number words for 6, 7, 8 and 9, that is they do not decompose 6 into 5 + 1, 7 into 5 + 2, etc. But their method of *writing* numbers clearly shows that they had a 5–20-system. They represented 5 by a bar and 1 by a dot. Numbers below 20 were represented by bars and dots, e.g. 17 by three bars and two dots: 17

For 20 the Mayas had a special sign: (see Figure 7).

This sign with, for example, two or three dots meant $2 \times 20 = 40$, $3 \times 20 = 60$, and with one or two bars it meant $5 \times 20 = 100$, $10 \times 20 = 200$ and so on.

From this it may be concluded that the Mayas once had a complete 5–20-system just as did their neighbours in Mexico.

The analogy of climbing a staircase helps one understand the different counting systems. Thus, 5-counting can be compared with climbing a staircase consisting of small steps of height 1 and larger steps of height 5 (see Figure 8). After having climbed five small steps, a platform of height 5 is reached. On this platform, another staircase of small steps is placed and one counts $5 + 1, 5 + 2, 5 + 3, 5 + 4$. Next, one reaches another platform of height 10 and so on.

The 5–10-system is illustrated by a staircase in which 5 small steps lead up to a platform of height 5 and two larger steps of 5 to a platform of height 10 (see Figure 9).

FIGURE 7   Mayan numbers on stone fragment found in British Honduras (British Museum)

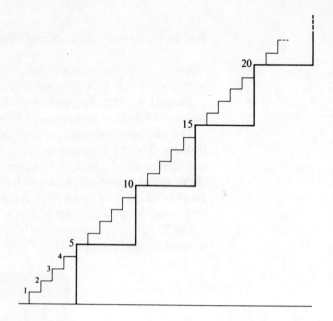

FIGURE 8　Staircase of 5-count

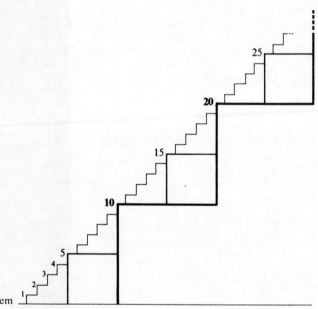

FIGURE 9　Staircase of 5–10-system

In the same way the 5–20-system is represented by a staircase built out of small steps of height 1, larger steps of height 5 and big steps of height 20 (see Figure 10).

A particularly interesting variant of the system is that of the Ainu, a small group of hunters and sea-food gatherers living in the northern islands of Japan, mainly in Sakhalin. They belong to the white race but their language is unrelated to any other known speech.

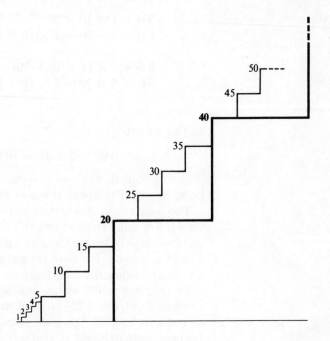

FIGURE 10   Staircase of 5–20-system (unit steps have been omitted after 5)

The Ainu have just seven number words from which all others are combined, namely

1 *shi-ne*	4 *ine*
2 *tu*	5 *ashik-ne*, meaning 'hand'
3 *re*	10 *wan*, meaning 'two-sided' (two hands)
	20 *hot-ne*, derived from *hot* = 'the whole' and *ne* = 'to be'

The numbers from 6 to 9 are obtained by subtraction:

$$6 = 10 - 4 \text{ } \textit{i-wan} \text{ (4 from 10)}$$
$$7 = 10 - 3 \text{ } \textit{ar-wan} \text{ (3 from 10)}$$
$$8 = 10 - 2 \text{ } \textit{tu-pesan} \text{ (2 steps down)}$$
$$9 = 10 - 1 \text{ } \textit{shine-pesan} \text{ (1 step down)}$$

The last two expressions show that the Ainu actually thought of counting as climbing a staircase.

Beyond 20, the counting goes on as follows:

$$30 = 2 \times 20 - 10 \text{ } \textit{wan e tu hotne}$$
$$40 = 2 \times 20 \qquad \textit{tu hotne}$$
$$50 = 3 \times 20 - 10 \text{ } \textit{wan e re hotne}$$
$$60 = 3 \times 20 \qquad \textit{re hotne}$$
$$70 = 4 \times 20 - 10 \text{ } \textit{wan e ine hotne}$$
$$80 = 4 \times 20 \qquad \textit{ine hotne}$$
$$90 = 5 \times 20 - 10 \text{ } \textit{wan e ashikne hotne}$$
$$100 = 5 \times 20 \qquad \textit{ashikne hotne}$$
$$110 = 6 \times 20 - 10 \text{ } \textit{wan e iwan hotne}$$
$$120 = 6 \times 20 \qquad \textit{iwan hotne}$$

$$200 = 1 \times 10 \times 20 \qquad \textit{shine wan hotne}$$
$$300 = 5 \times 20 + 1 \times 10 \times 20 \qquad \textit{ashikne hot-ikashama shine}$$
$$\textit{wan hotne}$$
$$400 = 2 \times (1 \times 10 \times 20) \qquad \textit{tu-shine wan hotne}$$
$$500 = 5 \times 20 + 2 \times (1 \times 10 \times 20) \ \textit{ashikne hot-ikashama-tu-}$$
$$\textit{shine wan hotne}$$

and so on until

$$1000 = 5 \times (1 \times 10 \times 20) \ \textit{ashikne-shine wan hotne}$$

This very complicated way of forming number words is described by Menninger as being 'a truly Cyclopean tower of numbers'[9].

The 5–20-system has certain marked differences from our system. As Figure 10 shows, five small steps of height 1 lead up to an intermediate platform of height 5 and four steps of height 5 lead to a platform of height 20. The next factors are again 5 and 4 and so on. It is found in many parts of the world (Figure 13). Later we shall see that it is closely connected with *finger-and-toe counting*.

By referring to the appropriate staircases, our decimal counting system can be compared with the 5–20-system. Our decimal system is a perfectly regular counting system with base 10. Ten small steps of height 1 lead up to a platform of height 10, ten large steps of height 10 lead up to a still larger platform of height 100, and so on. The next step is always 10 times the preceding one. The 5–20-system is different. Its fundamental factors are alternatively 5 and 4. The heights of the successive platforms are:

$$5, 20, 5 \times 20, 20^2, 5 \times 20^2, 20^3, \ldots$$

It can thus be called an *alternating system* because of its alternating factors. The first step 5 was chosen because we have five fingers on each hand and 5 toes on each foot. We have two hands and two feet, making 4 groups of 5, all in all a total of 20 fingers and toes.

Another alternating system is the *10–60-system*. The reader will not find it in Figure 13 because it is no longer used. It was used by the Sumerians, who reached a high level of civilisation in Southern Babylonia (near the Persian Gulf, where the rivers Euphrates and Tigris come together) in the third millennium BC. Their number words below 60 were formed as follows:

1 *ash* or *gesh*
2 *min*
3 *esh*
4 *limmu*
5 *ia*
6 *ash* (from *ia-ash* = 5 + 1)
7 *imin* (from *ia-min* = 5 + 2)
8 *ussu*
9 *ilimmu* (from *ia-limmu* = 5 + 4)
10 *u*
20 *nish*
30 *ushu* (from *esh-u* = 3 × 10)

40 *nin* (from *ni-min* = 20 × 2)
50 *ninu* (from *ni-min-u* = 20 × 2 + 10)

Up to this point there is a normal 10-system with traces of 5–20-count – nothing unusual. However, at 60 a completely new phenomenon occurs: the number 60 is considered as a new unit and is denoted by the same word *gesh*, which can also have the meaning 'one'. All higher steps are multiples of 60:

$$
\begin{array}{ll}
60 & gesh \\
2 \times 60 & gesh\text{-}min \\
3 \times 60 & gesh\text{-}esh \\
10 \times 60 & gesh\text{-}u \\
60 \times 60 & shar \\
10 \times 60^2 & shar\text{-}u \\
60^3 & shar\text{-}gal
\end{array}
$$

The word *shar-gal* means 'big-shar'. From the formation of this word we may conclude, with a certain degree of probability, that *shar* = 60^2 was at one time a limit of counting and that at a later time the limit was extended to 60^3.

Considering the system as a whole it becomes clear that it is a mixture of two quite different systems: an older 5–20-system, which was retained for numbers below 60, and a more recent (but still very old) 10–60-system (see Figure 11), in which the stepping-stones were

$$1, 10, 60, 10 \times 60, 60^2, 10 \times 60^2, 60^3$$

In the written numerals of the Sumerians there is no trace of a 5–20-system. One is bound to conclude that the 5–20-system was developed as a system of spoken number words before the Sumerians learned to write numbers. On the other hand it is quite possible that the 10–60-system was first developed as a system of writing numbers and that the number words for 60, 10 × 60, etc. were formed under the influence of the writing system.

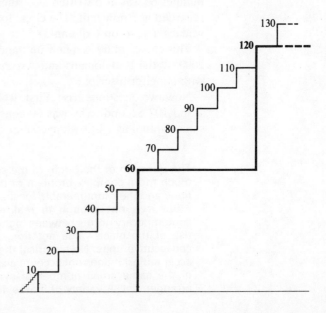

FIGURE 11 Staircase of 10–60-system (unit steps have been omitted after 10)

In any case the written numerals of the Sumerians correspond exactly to the number words just mentioned. The signs were impressed in a clay tablet by means of a stylus. Originally, a round stylus was used for writing numbers. If the stylus was impressed vertically a circular impression resulted, which meant 10, and if it was put in a slanted position a semicircular sign was obtained which meant 1 (see Figure 5). A larger semicircle, drawn by a sharp stylus, meant 60 and a larger circle meant 60^2. By combining the symbols for 10 and 60 the number 60×10 was written and, by combining the signs 60 and 60^2, 60^3 could be written (see column 3 of Figure 12).

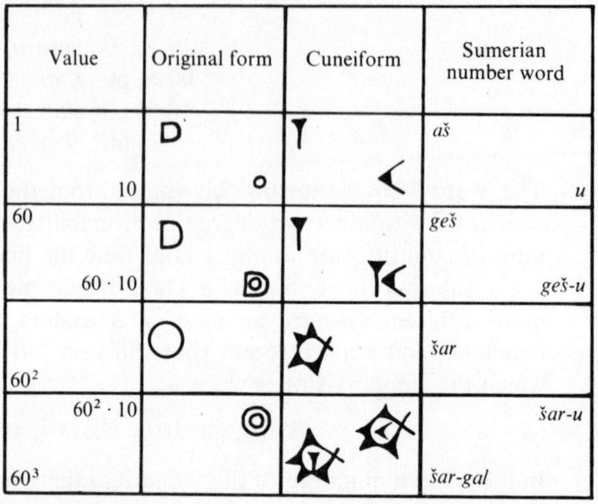

Value	Original form	Cuneiform	Sumerian number word
1	D	Y	aš
10	o	<	u
60	D	Y	geš
60 · 10	ⅅ	Ⱄ	geš-u
60^2	O	✡	šar
60^2 · 10	◎	✦	šar-u
60^3		✦✦	šar-gal

FIGURE 12   Early Sumerian numerals and number words (third millennium BC)

Later on the number 1 was written as a wedge and 10 as an L-shaped wedge. The number 60 was also written as a single wedge, thus expressing the idea that 60 was regarded as a new unit. The circle for 60^2 was replaced by a square formed by four wedges and so on (column 3).

This choice of 60 as a new unit and base of the system is unique in history. After 2000 BC the Babylonians and Assyrians took it over from the Sumerians together with cuneiform script.

Now two questions arise. First, what led to the choice of 60 as the next stage after 1 and 10? Second, why was 60 considered as a 'big unit' and represented by the same symbol as 1? O. Neugebauer discusses these problems as follows:

The question of the origin of the sexagismal system is inextricably related to the much more complex problem of the history of many concurrent numerical notations and their innumerable local and chronological variations.

But it is not enough to realize that the 60-division is only one of several contemporary norms between higher and lower units. The essential point lies in the use of the place value notation, regardless of the value of the ratio between consecutive units. No historical theory of the origin of the sexagesimal system is acceptable if it does not account also for this extraordinary feature, namely, the use of the same small number of symbols for different values, depending on the arrangement. A variety of 'bases' is well known from number words and number writing all over the world. The place value notation, however, is the most striking feature of the Babylonian system. A problem of this kind cannot be solved by speculation, but only by a systematic analysis of the written documents.[10]

Neugebauer points out that the overwhelming majority of cuneiform texts are concerned with economic matters, and not with religion, magic or number mysticism. He continues:

> In economic texts units of weight, measuring silver, were of primary importance. These units seem to have been arranged from early times in a ratio of 60 to 1 for the main units 'mana' (the Greek μινα 'mina') and shekel. Though the details of this process cannot be described accurately, it is not surprising to see this same ratio applied to other units and then to numbers in general. In other words, any sixtieth could have been called a shekel because of the familiar meaning of the concept in all financial transactions. Thus the 'sexagesimal' order eventually became the main numerical system and with it the place value writing derived from the use of bigger and smaller signs. The decimal substratum, however, always remained visible for all numbers up to 60. Similarly, other systems of units were never completely extinguished. Only the purely mathematical texts, which we find well represented about 1500 years after the beginning of writing, have fully utilized the great advantage of a consistent sexagesimal place value notation. Again, 1000 years later, this method became the essential tool in the development of a mathematical astronomy, whence it spread to the Greeks and then to the Hindus, who contributed the final step, namely, the use of the place value notation also for the smaller decimal units. It is this system that we use today.

Nothing is more natural, when one monetary unit is 60 times another, than denoting the larger unit again by 1 and its multiples by the ordinary symbols for 1, 2, ..., 10, ... We all do it every day. 'Two-fifty' just means 2 pounds and 50 pence, or 2 dollars and 50 cents, as the case may be.

The Sumerian system of numerals will be discussed in greater detail in Chapter 4, and it will be seen how it quite naturally led to a place-value system for integers and fractions with the base 60.

Figure 13 shows the geographical distribution of the various counting systems. It is a simplified version of Map XII in W. Schmidt's *Sprachfamilien und Sprachenkreise der Erde*, Heidelberg, 1926 (*Atlas*). Schmidt distinguishes ten counting methods from systemless counting to the pure decimal system (10-count). Figure 13 is a simplified version of the map and it distinguishes only between

2-count
4-count
20- and 5–20-count
10- and 5–10-count

Schmidt also has small regions of 6-count and 40-count, and white regions about which he says nothing. These regions have been left white in the figure.

It is often difficult to say whether a counting-system is 5–10- or pure 10-, or whether it is 5–20- or pure 20-. The characteristic property of a 5–10- and 5–20-system is that 6, 7, 8 and 9 are pronounced as $5 + 1, 5 + 2, 5 + 3$ and $5 + 4$. In many cases, only traces of these decompositions are found. In those cases, where it

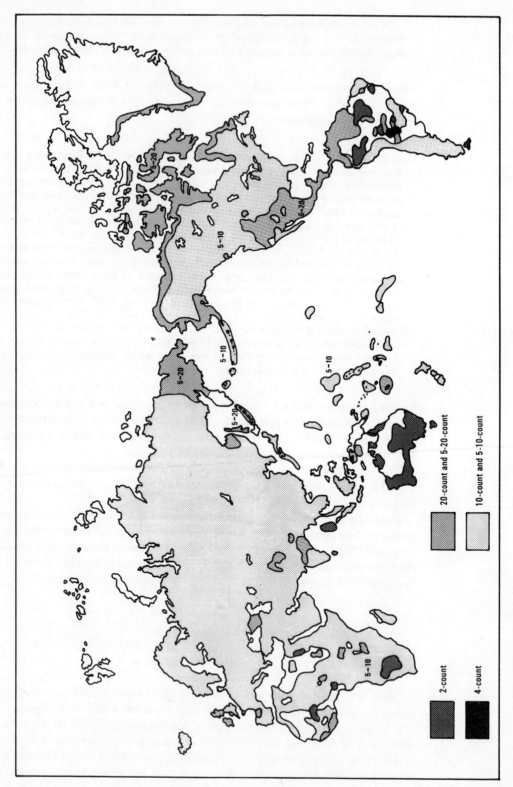

FIGURE 13 Distribution of counting systems

is certain that the system is 5–10- or 5–20-, the symbol '5–10' or '5–20' has been inserted.

For Europe and Asia, Figure 13 maps the counting systems as they are today. In the other three continents only the counting methods of the aborigines are shown on the map and not the decimal system used by the white population.

Leaving the 4-system and the 5–10-system out of account, one is left with three principal systems, namely:

> the pure 2-system,
> the 20-system (5–20- and pure 20-),
> the 10-system.

The pure 2-system is very primitive: it cannot well be extended beyond 10. For this reason, it was superseded in most parts of the world by the other (better) systems. The 2-system remained only in isolated parts of the Southern hemisphere.

The pure 10-system is the most efficient of all; it therefore tends to spread out over the world. This tendency can be observed in the time of the Celts and Romans as well as in the last five centuries. In the Middle Ages, English was spoken only in Britain, now it is the principal language of, for example, North America and Australia. The Romance languages are now spoken in the whole of Southern Europe and South America and with these languages came the decimal system. The Chinese also have the decimal system and their civilisation too spread out over a large area. So the spread of the decimal system is easy to explain. Today it covers by far the largest part of the surface of the earth.

The history of the 5–20-system is much more complicated and largely unknown. It was fully developed by the Aztecs and Mayas, in Mexico and Yucatan, and from this region it spread to the north and the south, replacing the less developed 5–10-count in the north and the primitive 2-count in the south. We find 5–20-count also among Eskimos. They all speak dialects of one single language and their racial features point to an Asiatic origin. Hence, most probably, the 5–20-system was brought by the Eskimos from a centre in north-east Asia to the north coast of North America and to Greenland.

In Western Europe, the 20-system (perhaps originally a 5–20-system) once prevailed in a large area, including Spain, France, Great Britain, Ireland and some Scandinavian countries. When the Celtic and Germanic languages spread towards the West, the 20-system was largely replaced by the 10-system, but it maintained itself in the Basque language and traces of 20-count remained also in Irish, Scottish, French, Danish and other languages.

All in all it seems that the 5–20-system has an intermediate position between the more primitive pure 2-system and the more efficient 10-system. In regions where the 10-system is fully developed up to 1000 or even beyond, it tends to spread out to replace the 5–20-system. On the other hand, in regions where the 5–20-system is fully developed (as in Middle America) this system is able to penetrate not only into regions of 2-count but even into regions of 10-count.

In a rather complicated argument, which will not be considered here, Seidenberg suggests that the 5–10-system may be simply a cross between a quinariless 10-count and the 5–20-system. For this reason, the 5–10-system has not been included amongst the three principal systems considered above. It has been left out of account along with the 4-system.

Remnants of ancient counting

Modern Western languages are Indo-European in origin, and the counting is purely decimal. Yet relics of a 20-system remain. The French, for example, pronounce 80 and 90 as *quatre-vingt* and *quatre-vingt-dix*, that is as 4 × 20 and 4 × 20 + 10. They also pronounce 70 as 60 + 10. The English language has the expression 'three-score' for 60. In the Authorised Version of the Bible is found:

> The days of our years are three-score years and ten; and if by reason of strength they be fourscore years, yet is their strength labour and sorrow; for it is soon cut off, and we fly away. (Psalm 90:10)

Shakespeare also counts by scores.

> Why, this boy will carry a letter twenty miles as easy as a cannon will shoot point-blank twelve score. (*Merry Wives of Windsor*, Act II, Scene 2)

Menninger draws particular attention to examples from the everyday language of the common people. He writes:

> It is difficult for an illiterate peasant to tell how many years he has lived; the Piedmontese, for instance, would answer that he carries *quat borla*, '4 large loads', on his bent back, meaning 4 × 20 = 80 years. This is far easier for him to visualize; he does not climb the steps of his years age, 77, 78, 79, 80, but remains standing on the ground and, like a mule, takes up the 'loads' of his years grouped in bundles of 20. The Sicilian dialect has no special word for the 20-group, so that in Sicily the peasant says *tri vintini et deci*, '3 twenties and 10', for 70 years; thus he uses the number word *vinti*. In some regions of Spain the monetary units (1 *duro* = 20 *reales*) are used in counting, so that a peasant there would reply that he is '3 duros and 10 reales' old.[11]

Another example of counting money in twenties can be found in Portugal, where the word *vintens*, 'twenties', is in use. Sometimes it is not actually pronounced: *dois cinco* is an abbreviation for *dois vintens e cinco* (2 × 20 + 5 = 45) and *seis menos cinco* means 'six twenties less five' (= 115).

Probably the oldest documentary evidence is an inventory of an English monastery, written about AD 1050, which contains expressions from vernacular language together with their Latin translations:

> V *scora scoep*   *quinquies viginti oves*   (5 × 20 sheep)
> VIII *scora oecer*   *octies viginti agri*   (8 × 20 acres)

In Sweden, farmers often say *fyra sneser* (four *snes*) when they mean eighty. In Holland some peasants count eggs by the *snees*, a *snees eieren* meaning 20 eggs. The word *snes* or *snees* (German *Schneise*) has the original meaning of a rod or twig used to line things up in a row. A Baltic monastic register from 1186 includes in its annual income 24 *snesas anguillarum* = 24 × 20 eels.

Another example of counting by twenties is found in the Danish language. Up to 40 the counting is decimal:

> 10 *ti*
> 20 *tyve*
> 30 *tredive*
> 40 *fyrretyve*

but 60 and 80 are considered as multiples of 20:

$$60 = 3 \times 20 \; \textit{tre-sinds-tyve}$$
$$80 = 4 \times 20 \; \textit{fir-sinds-tyve}$$

The expressions for 50, 70 and 90 are very curious:

$$50 = 2\tfrac{1}{2} \times 20 \; \textit{halv-tre-sindstyve}$$
$$70 = 3\tfrac{1}{2} \times 20 \; \textit{halv-fjerd-sindstyve}$$
$$90 = 4\tfrac{1}{2} \times 20 \; \textit{halv-fem-sindstyve}$$

In the last expression, *half-fem*, literally 'half-five', means $4\tfrac{1}{2}$, just as *half fünf* in German means 'half past four'. So the numbers 50, 70 and 90 were considered as half-integer multiples of 20.

In Celtic languages (see Figure 14) traces of popular 20-count are again found. In

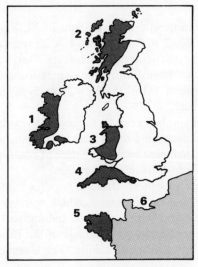

FIGURE 14   Areas in which Celtic was and is spoken: 1 Irish; 2 Gaelic in Scotland; 3 Welsh; 4 Cornish (extinct); 5 Breton; 6 Gallic (extinct)

old Irish there is a regular decimal system:

10 *deich*	60 *sesca*
20 *fiche*	70 *sechtmoga*
30 *tricha*	80 *ochtmoga*
40 *cethorcha*	90 *nocha*
50 *coica*	100 *cet*

But in popular speech there were expressions such as

deich mnaa a secht fichit

which means '10 women and seven twenties', i.e. 150 women.

This popular tendency to count by twenties has led to a regular 20-system in New Irish and in Scottish Gaelic. In New Irish 40 is pronounced as *da fichit* (2 × 20), 60 as *tri fichit* (3 × 20) and so on.

A similar development can be found in Welsh, Cornish and Breton, which are all closely related languages. The Welsh number words are:

1 *un*	11 *un-deg-un*
2 *dau (f. dwy)*	12 *deuddeg* or *un-deg-dau*
3 *tri (f. tair)*	13 *un-deg-tri*
4 *pedwar (f. pedair)*	14 *un-deg-pedwar*
5 *pump*	. . .
6 *chwech*	20 *ugain*
7 *saith*	30 *tri-deg* or *deg-ar-hugain*
8 *wyth*	40 *pedwar-deg* or *deugain*
9 *naw*	60 *chwe-deg* or *trigain*
10 *deg*	80 *wyth-deg* or *pedwar-ugain*

Here in the alternative forms of 30, 40, 60, 80 are the number-words equivalent to $10 + 20$, 2×20, 3×20, 4×20 respectively.

It is particularly interesting that, although French is a Romance language (derived from Latin), the expression *quatre-vingt* is not itself derived from a Latin equivalent. Also, the New Irish *da fichit* is not derived from an equivalent form in Old Irish.

A hypothesis can be suggested to explain why forms like *quatre-vingt* and *da fichit* are present in French and New Irish, although they are absent in Latin and Old Irish, namely that the practice of counting by twenties derives from a more ancient non-Indo-European language.

This hypothesis that the popular tendency to count by twenties is derived from a more ancient language, which was spoken in Western Europe before the Celtic period and which did not belong to the Indo-European language family, finds confirmation in the language of the Basques (see Figure 15) which is not related to the Indo-European family. The Basque words for 1 to 9 and 10 to 90 are:

FIGURE 15   Area of the Basques

1 *bat*	10 *hamar*
2 *biga*	20 *hogoi*
3 *hirur*	30 = 20 + 10 *hogoi-ta-hamar*
4 *laur*	40 = 2 × 20 *berrogoi*
5 *bortz*	50 = 40 + 10 *berrogoi-ta-hamar*
6 *sei*	60 = 3 × 20 *hirur-hogoi*
7 *zazpi*	70 = 60 + 10 *hirur-hogoi-ta-hamar*
8 *zortzi*	80 = 4 × 20 *laur-hogoi*
9 *bederatzi*	90 = 80 + 10 *laur-hogoi-ta-hamar*

This evidence, together with evidence from certain other non-Indo-European languages, leads to the conclusion that in Western Europe before the expansion of Indo-European languages a 20-system of counting was in use. Afterwards the number words were replaced by Indo-European number-words like *quatre* and *vingt* but the tradition to count by multiples of 20 was retained in many places.

It is not known whether the Basque 20-system originally was a 5–20-system. It seems possible that the word *sei* for 6 was taken over from a Romance language and that 6 was originally pronounced as 5 + 1. Another trace of Romance influence is the Basque word *milla* for 1000.

Finger and body counting

Finger counting

Finger counting is a very natural procedure: one might assume that every group could discover it for itself. Yet examination of the details of finger counting shows that there are traditions in finger counting which were passed on from one group to another.

The number words of the Australian Chaap Wurrong tribe in Victoria from 4 to 50 have the following meaning:

4 'two and two'
5 'one hand'
6 'one finger, hand'
7 'one hand and two'
8 'one hand and three'
9 'one hand and two and two'
10 'two hands'
11 to 19 ten plus the symbol for 1 to 9
20 'one twenty'
30 'one twenty and ten'
40 'two twenties'
50 'two twenties and ten'

The counting systems 2-count, 5-count and 20-count can be detected in the Chaap Wurrong number words.

The 5–20-system is very often associated with finger counting or with counting on fingers and toes. Seidenberg writes:

> Returning to Central Brazil, we note that of the dozen or so groups that von den Steinen met, only two, the Bakairi and the Bororo, use the 2-system. Rather, an independent word exists for 5 [i.e. not composed of 2 + 2 + 1]; then 6, 7, 8, 9 are constructed from 1, 2, 3, 4 with some modification to show that the speaker is in the second pentad. These tribes use their fingers and toes in counting, and in most, though not all of them, the word for 5 is related etymologically to 'hand'. 'Ten' is with some of them a new word, with others it contains (etymologically) 'the hands'. Then for 11, 12, 13, 14 one modifies 1, 2, 3, 4 with an addition referring to a 'foot'.
>
> .   .   .   .
>
> In all, with one exception, 'feet' occurs in the word 'twenty'.[12]

It is clear that here at least there is a basic fragment of a 5–20-system, and that this 5–20-system is closely connected with *finger-and-toe counting*.

The same phenomenon occurs in New Guinea. Menninger writes:

> In translating the Bible for one tribe of Papuans, the passage (John 5:5): 'And a certain man was there, which had an infirmity thirty and eight years' had to be expressed as 'a man lay ill one man (20), both sides (10), 5 and 3'. Even more picturesque is the expression for the number 99 in British New Guinea: 'four men die (80), two hands come to an end (10), one foot ends (5) and 4'.[13]

Both cases are clearly 5–20-systems, and the connection with finger-and-toe counting is equally clear.

The examples from Australia, South America and New Guinea, are by no means the only ones. In many other cases Seidenberg has found that 5-count was combined with finger counting, and 5–20-count with finger-toe counting. The only reason why people all over the world should have counting systems based on the numbers 5 and 20 would seem to be that there are 5 fingers on each of 2 hands and 5 toes on each of 2 feet.

When considering different methods of finger counting some curious evidence comes to light. There are large connected regions in Africa, Southern Asia, Australia and America in which counting always starts with the left little finger. In other regions it starts with the right little finger (see Figure 16). In the middle part of Africa and in South America there are regions in which the right index finger is 1.

There are four connected regions in which counting starts with the thumb, namely one region in Western Europe, one in New Guinea and Melanesia, one in Middle and South America and one in Greenland. These regions are indicated in Figure 16. It is interesting to note that in all these regions the 20-count is also found. Seidenberg writes:

> These [distributions] may at first glance seem random, but it has been my observation that in almost every case in which the thumb is 1, the system is vigesimal.[14]

Seidenberg uses this coincidence as an argument in favour of his hypothesis that every systematic counting system, such as 5–20-count, was invented only once. This seems to be a good probability argument. In fact, if the 5–20-system and thumb counting had been invented independently in (say) three or four regions, the

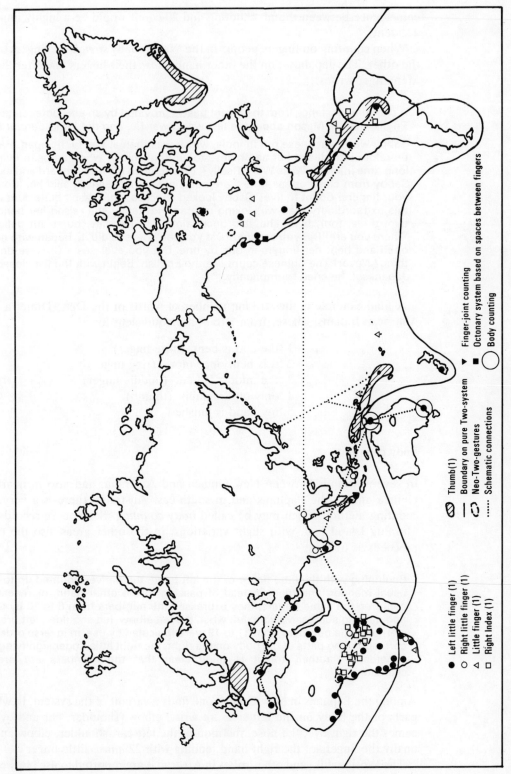

FIGURE 16 Finger-counting practice (by kind permission of Professor A. Seidenberg)

coincidence between thumb counting and 20-count would be a highly improbable accident.

When counting on fingers people in the West usually *stretch* the fingers one after the other. The Japanese, on the other hand, *close* their fingers one after the other. Menninger writes:

> A charming instance from the recent past is provided by an anecdote related by the Englishman R. Mason about the last world war (*... and the Wind Cannot Read*):

> Sabby was a Japanese girl in India, which was then at war with Japan. Her friend therefore introduced her as Chinese to an Englishman who had been living for a long time in India: 'Miss Wei'. 'Really?' He stretched his face forward and examined Sabby from close up, as if he were nearsighted. 'Nonsense', said he. 'Count with your fingers! Count to five!' Sabby looked shocked; she wasn't quite sure whether this extraordinary man was joking or mad. Hesitantly she raised her hand: 'One, two, three, four, five,' she said uncertainly. Mr. Headley, burst out delightedly: 'There you are! Did you see that? Did you see how she did it? Began with her hand open and bent her fingers in one by one. Did you ever see a Chinese do such a thing? Never! The Chinese count like the English. Begin with the fist closed. She's Japanese!' he cried triumphantly.[15]

A final example is the striking number of words of the Dene-Dinje, a tribe of American Indians. These, from 1 to 5, are equivalent to:

1 'the end is bent' (little finger)
2 'it is bent once more' (ring finger)
3 'the middle is bent' (middle finger)
4 'only one remains' (thumb)
5 'my hand is finished'

Body counting

In several islands between New Guinea and Australia, and also in nearby New Guinea among the Papuans and in south-east Australia, there is a very curious counting method which may be called *body counting*. Seidenberg records that in Muralug Island and, with slight variations, in the other areas also the counting proceeds as follows:

> Beginning with the little finger of the left hand, the natives counted up to 5 in the usual manner, and then, instead of passing to the other hand, or repeating the count on the same fingers, they expressed the numbers from 6 to 10 by touching and naming successively the left wrist, the left elbow, left shoulder, left breast, and sternum. The numbers from 11 to 19 were indicated by use, in inverse order, of the corresponding parts of the body on the right, the right little finger signifying 19 – or 'nineteenth', rather, for it is fairly clear that the numbers are ordinal in character.[16]

Among the Papuans in New Guinea one finds a variant of the system, in which the parts of the body on the left side are wrist, elbow, shoulder, ear and eye. Next comes the right eye, the nose, the mouth, the left ear, shoulder, elbow and wrist, finally the fingers of the right hand, ending with 22 *anusi* 'little finger'.

The Wotjobaluk, and other tribes in Australia, are reported to use body counting only for counting days of a journey. It is curious that body counting occurs only in areas of 2-count (see Figure 3).

Seidenberg attempts to solve the apparent paradox that the more complicated system of body counting coincides with the simplest form of number words, 2-count, by suggesting that finger counting is the residue of a complex system of parcelling out various parts of the body to various gods.

Tallies and knots

Tallies

The history of numbers, in its widest sense, is concerned with representations of numbers by knots and by notches carved on pieces of wood or bone, as well as with specific symbols engraved on wood, stone or metal, or written on clay, papyrus, parchment, etc.

The earliest known numerals were written on clay by the Sumerians around 3000 BC. However, there is evidence of *tallying* (carving notches on bone or wood) going back more than thirty thousand years.

Finger and body counting provide no permanent record of any calculation although they do share the visual quality of written numbers – a quality which spoken words do not possess. Finger and body counting can thus be thought of as occupying an intermediate position between spoken numbers and written numbers.

Today one invariably thinks of written numbers in terms of systems of special number symbols. But if such symbols had not been invented would there be any way of making a permanent record of calculations? Menninger takes up this idea:

> Suppose someone were to do away with all numerals, even the Roman? Then we would have some idea of the state of written numbers in the early Middle Ages. To be sure, Roman numerals found their way into the monastic cloisters along with other aspects of the culture inherited from ancient Rome; of course, Roman numerals were later brought by the students from these monastic schools into the world of literate people, of merchants, of scribes in chanceries. But the farmer, who never went to a monastery school, also had to keep track of his crops and livestock and harvests. The peasant could also be a creditor or a debtor. Where did he get the numerals to reckon with?[17]

Menninger's answer is that he used carvings on a tally stick.

The oldest example of tallying so far discovered dates back to paleolithic times. It is a wolf bone found in 1937 in Vestonice (Moravia), shown in Figure 17.

The excavations were conducted by Dr Karl Absolon and were particularly interesting because, in addition to the wolf bone, they unearthed a sculptured ivory head of a woman, the earliest human portrait by a paleolithic artist so far discovered. Dr Absolon comments on the wolf bone as follows:

FIGURE
17 Paleolithic wolf bone found in Moravia in 1937 (Maravské Museum, Czechoslovakia)

> An exceedingly valuable find was a radius of a young wolf, 18 cm (about 7 in.) long, engraved with fifty-five deeply incised notches. First come twenty-five notches, more or less equal in length, in groups of five, followed by a single notch twice as long, which terminates the series; then, starting from the next notch, also twice as long, a new series runs up to thirty. We know several of these Diluvial objects with

markings denoting the conception of five. This number corresponds to the five fingers on each hand, which Diluvial man constantly had before his eyes and which no doubt led him to think in terms of numbers. We have here direct proofs of reckoning by fossil man.[18]

Another interesting find is a bone tool handle of the later stone age discovered at Ishango on Lake Edward, Zaire (see Figure 18). The discoverer, Dr Jean de Heinzelin, suggests that it was probably the handle of a tool used for engraving or writing. The bone is described and discussed as follows:

There are three separate columns, each consisting of sets of notches arranged in distinct patterns. One column has four groups composed of eleven, thirteen, seventeen and nineteen notches; these are the prime numbers between ten and twenty. In another column the groups consist of eleven, twenty-one, nineteen and nine notches, in that order. The pattern here may be 10 + 1, 20 + 1, 20 − 1, and 10 − 1. The third column has the notches arranged in eight groups, in the following order: 3, 6, 4, 8, 10, 5, 5, 7. The 3 and 6 are close together, followed by a space, then the 4 and 8, also close together, then another space, followed by 10 and two 5's. This arrangement seems to be related to the operation of doubling. De Heinzelin

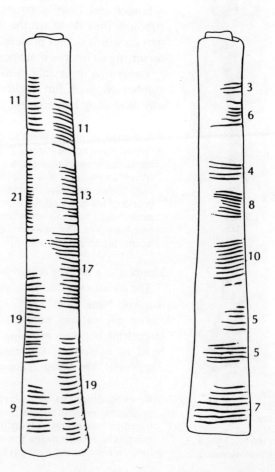

FIGURE 18 Schematic drawing of the Ishango bone showing the arrangement of notches (Prindle, Weber & Schmidt, Boston, Mass.)

concludes that the bone may have been the artifact of a people who used a number system based on ten, and who were also familiar with prime numbers and the operation of duplication.[19]

The use of tally sticks seems to have been universal. Menninger explains this as follows:

Paper, which was a Chinese invention, first came to Germany in the 14th century. In the beginning it was too expensive for common use; like parchment in the monasteries, it was used only for the most important documents. Moreover great skill was required for writing on it. The common people's 'paper' was wood, its writing instrument was a stylus or sharp knife and its letters were notches or grooves. For this reason, we find some form of tally used everywhere.[20]

Figure 19 shows examples of tallying from the Fiji Islands and from the Philippines. Both show clear evidence of *grouping*.

On the lower part of the Fiji Island club there is a group of nine notches followed by a larger notch. On the Philippine sword are groups of three. Thus the 'language' of tally sticks was by no means universal. Menninger makes this point:

Who knows what numbers are signified by the notches on the war club from the Fiji Islands? Perhaps a member of the tribe which carved the club could still read them,

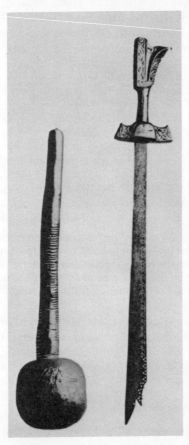

FIGURE 19   Fiji island club with number notches, 40 cm (Museum für Volkerkunde, Stadt-archiv, Frankfurt am Main) and Philippine sword, on whose blade the owner has recorded the number of his victims with silver nails inlaid in groups of three, 94 cm (Linden-Museum, Stuttgart)

just as many Swiss can identify a particular hamlet or valley from the carvings on a notched stick. But their significance is always personal and valid only within a limited area: they are folk-signs, number symbols devised by simple people themselves to meet their everyday needs. They are not learned and taught in school as part of the general culture.[21]

The Bakairi tribesman of Central Brazil counts by pairs up to six.[22] This example of primitive grouping shows an organisation of counting which involves both ordering and grouping. Menninger comments:

Up to 6 objects the Bakairi *groups* by pairs; from there on he merely *orders* them. Which was then the earlier stage? Ordering, of course: without first ordering, there can be no grouping. The Bakairi stands, as it were, with one foot on the lower step of ordering and diffidently places the other on the next higher step of grouping by twos: another picturesque instance of how numbers first became clear to primitive man through the use of a supplementary quantity.

But why grouping? Let us ask ourselves why do we always divide a bundle of postal cards by twos? Why do we break up a pile of coins into little heaps of five? To count them better! 'Better' means more clearly, more confidently, and therefore more quickly.[23]

The Latin word for tally is *talea*, which means 'cut twig' or 'stave'. In Italy the tally was called *taglia* or *tessera*, in Spain *tarja*, in France *taille*. Of course, the English word *tally* is derived from French *taille*. The Swedish word is *karvstock*. The corresponding Dutch word *kerfstok* is seldom used in its literal meaning, but has become a synonym of 'bad conscience'. If a Dutchman says

> *Hy heeft iets op zyn kerfstok*
> (literally: 'He has something on his tally-stick')

he means: 'He has done something wrong'. The German word *Kerbholz* has the same meaning: A German would say

> *Er hat etwas auf dem Kerbholz*

A contemporary of the Flemish painter Pieter Brueghel the elder (*d*. 1569) records the following anecdote:

While he was in Antwerp, he lived with a young girl whom he would have married if she had not had the unfortunate habit of constantly telling lies. He made a pact with her that for each lie she told he would cut a notch in a piece of wood; he took a good long stick, and when the stick was filled with notches, there could be no longer any question of marriage. This happened before long.[24]

One of the main purposes of the tally stick was to indicate debts. In Schiller's play *Wallenstein's Camp* a barmaid presents a bottle to the company, saying

> *Das kommt nicht aufs Kerbholz. Ich geb es gern. Gute Verrichtung, meine Herrn!*

which means, freely translated:

> 'This doesn't go on the tally. I give it free. Success in battle may you see.'

An Italian saying goes:

> *Come è bello vivere in questo regno:*
> *Si mangia, si beve e si segna*
> *Tutto su un pezzo di legna!*

which means:

> 'How pleasant it is to live in this country:
> One eats, one drinks and one notes
> It all upon a piece of wood!'

There are many German synonyms of *Kerbholz*, for instance:

> *Dagstock* (day stick)
> *Knüppel* (cudgel)
> *Span* (chip)
> *Raitholz* or *Rechenholz* (reckoning stick)

In Swiss dialects are the words:

> *Alpscheit* (Alp log)
> *Tessele* (from Italian *tessera*)
> *Beile* (probably derived from mediaeval Latin *pagella*, 'scale').

The English word *score* is derived from the Old Saxon verb *sceran*, meaning 'to shear' or 'to cut'. The verb *to score* means 'to cut' or 'to keep count', and the noun *score* also has the meaning 'twenty'. This example shows, once more, how popular tallies once were.

The Latin word *putare* means literally 'to cut' (compare *amputare*, 'to cut off'). From this was derived *imputare*, 'to cut in', which took the figurative meaning 'to assign a debt to someone', obviously because debts were cut into a tally stick. *Computare* became the standard expression for 'to compute', 'to calculate', and *putare* acquired the meaning of 'to reckon', 'to think', 'to believe'.

How common bookkeeping on tally sticks once was, even in such a progressive commercial town as Frankfurt-am-Main in the fourteenth century, is shown by a quotation from the *Weaver's Statutes* of that town:

> It shall likewise be the rule for the entire guild that no calculator shall lend to one person the tallies that are owed to another.

Not all tally sticks were of the simple forms described so far. The inhabitants of the Nicobar Islands in the Indian Ocean often have to count the numbers of coconuts they have picked. Since a plain number stick would not suffice for this purpose because of the large numbers involved, they cut themselves a rod half a metre long from a bamboo stalk and split one of its ends into a number of lengths which remain connected like the brush of an old-fashioned broom. Then the number of coconuts is marked by notches carved in the split lengths at the end (see Figure 20).

The use of tally sticks as bills or records of obligations can be traced back to the time of the early Germanic tribes. A common feature of the laws of the Franci and Allemanni was the *festuca*, a rod that was exchanged between the parties to a

FIGURE 20   Number stick of split
bamboo used to count coconuts (by
courtesy of Vandenhoek & Ruprecht,
Göttingen)

transaction in matters of law, especially those involving promises of payment. According to Salic law the debtor had to turn over to his creditor his *festuca* on which his own identification symbol and the sum involved were marked. In this *fides facta per festucam*, this 'promise made by the *festuca*', the rod in question was nothing more than a notched stick on which the debt was tallied. Once the debt was paid, the tally stick was burned or else the sum was cut off, that is the creditor cut the notches off the stick: he 'made the tally smooth'.

Martin Luther, addressing a friend to whom he had not written for a very long time, began his letter:

I must cut off the tally stick, for I have not answered your letter for a long time.

A municipal document from the year 1588 says:

A woman who is a debtor is to pay 12 guilders per year, until the whole debt has been paid, and each time this payment is to be cut off from her tally stick.

'Milk sticks' (see Figure 21) were in use shortly before World War I in the Tavetsch Valley in Graubünden, Switzerland. These sticks were used to keep the 'books' of a whole village community. Each day the herdsman would carve a rod, some 15 to 20 cm long, which he coloured with red chalk; thus the uncoloured carving would stand out vividly. Then, for every peasant who owned cows in the common herd, an individual symbol was carved on the rod. Below each such symbol one sees an incised line, the individual peasant's 'account', on which the cowherd tallied off the amount of milk yielded by this peasant's cows. The milk was immediately used to make cheese and the sticks were used to settle the accounts between those who had made the cheese and those who owned the cows. The cheese makers and the cow owners were the same farmers: each day one of them went up into the high Alpine meadows to collect the milk and to make cheese. At the end of the day he would carry home the *stialas* showing how much milk he owed the cow owners. On a Sunday after church they would come together and settle the accounts.

In many countries, *double tally sticks* were in use until quite recently. They were made by cutting a long piece of wood lengthwise almost to the end. The part with the large end is called the 'stock' and the split-off portion is the 'inset'. In Vienna the two pieces were called *Manderl* and *Weiberl*, 'male' and 'female'. A beautiful Finnish double tally stick is shown in Figure 22.

The double tally stick shows the amount of money a debtor owes a creditor. The creditor usually keeps the stock while the debtor keeps the inset. The sums shown on both pieces are the same. Each time a partial payment is made the debtor inserts

FIGURE 21 'Milk sticks' (Ratisches Museum, Chur, Switzerland)

FIGURE 22 Finnish tally stick (National Museum of Finland, Helsinki)

his inset in the stock and notches are removed from both pieces at once. In this marvellously simple fashion the 'double bookkeeping' makes cheating impossible.

Double tally sticks were admitted as legal documents even as late as the eighteenth and nineteenth centuries. A Basel statute book of 1719 says:

> And if anyone is not good at settling his accounts by writing and reading, he shall be satisfied with crudely made tallies or tickets: Then if one party brings to court a tally or ticket as evidence of his debt, and the other party produces the corresponding or matching ticket or wooden tally, and they are found to be the same, credence shall be given to them and the amounts stated on them shall be acknowledged.

A similar statement can be found in Napoleon's book of civil law, the *Code civil* of 1804:

> The tally sticks which match their stocks have the force of contracts between persons who are accustomed to declare in this manner the deliveries they have made or received.

FIGURE 23   Chinese character for 'contract'

The word for 'contract' in Chinese is written by means of two characters at the top, one for a notched stick and one for a knife, and another at the bottom which means 'large' (see Figure 23). Thus a contract in Chinese is symbolised as 'a large tally stick'.

In English there are several popular expressions in which the word *tally* occurs. Thus, *the account does not tally* means 'the calculation is not correct' and *to keep tally with somebody* means to be in close relationship. Note also the French expression *acheter à la taille*, which means 'to buy on credit'.

An *Alpscheit* (literally: Alp log), from the Lötschental district in Switzerland, was a villager's certificate showing that he was entitled to let his cattle graze on the common pasture. Grooves and cuts of different lengths meant rights to pasture a certain number of cows or sheep. The peasants kept the inset pieces, or *Beitesseln*, as evidence of their individual pasture rights. The Alpscheit shown in Figure 24 bears the date 1752.

FIGURE 24   An *Alpscheit* with insets, 1752, on which the positive rights of village peasants are recorded. It is triangular in section, 130 cm long and 9 cm wide, weighs over 30 kg and has more than 70 cut-outs into which corresponding insets were fitted (by courtesy of Vandenhoek & Ruprecht, Göttingen)

The *Kapitaltesseln* (capital tesserae) of Visperterminen in Switzerland were the equivalent of bonds. The community had funds which it would lend out to individual farmers. As a promissory note for his debt the borrower would give the municipal government a tally stick with his personal mark carved on it and the amount of his debt marked on the opposite side. The tesserae were strung on a cord through the holes in their ends (see Figure 25).

Several hundred *Exchequer tallies*, notched wooden sticks used by the British Treasury, were discovered in 1909 during repairs to the Chapel of the Pyx, Westminster Abbey. Documents relating to their use and the remains of leather sacks in which the tallies had presumably been kept were also found. These British Exchequer tallies are unique in cultural history; some of them, dating from the thirteenth century, are shown in Figure 26.

Henry I (*r.* 1100–35) was the first king to introduce the wood tally system. Until his reign there had been no differentiation of function amongst the members of the Court or Council. Thus, there was no particular body nor any set of rules for the conduct of the financial affairs of the state. G. M. Trevelyan writes:

> The earliest step towards differentiation of function was taken in the reign of Henry I, when certain 'Barons of the Exchequer' evolved a procedure and an office of their own, inside the larger Court or Council, for the purpose of dealing with the most important of all the royal interests, the proper receipt of his multiform dues and

FIGURE 25   Capital tesserae; on the obverse side carved notches show the peasant's debt; on the reverse is his personal symbol (by courtesy of Vandenhoek & Ruprecht, Göttingen)

money payments from his sheriffs, feudatories, chartered boroughs and domain lands.[25]

The Court of the Exchequer took its name from a table in its central office covered by a woollen cloth on which a chequerboard pattern was drawn. Over this table the sheriffs of the provinces of England reported the amounts of taxes received or due, the *Calculator* laid out the amounts in *calculi* or counters on the chequered cloth and thus arrived at the final sum. Then the *Cutter* carved a *tally* to record the amount paid or owed to the Crown. The whole procedure was observed by members of the court and higher officials. The use of tally sticks and counters enabled everyone to follow the procedure without having to know how to read or write.

The Exchequer tallies were *double* (see p. 46), one piece being retained at the Exchequer and known as a *foil*, the other piece, know as the *stock*,[26] being issued in exchange for a sum of money paid in. They were sufficiently thick (usually about half an inch) for there to be writing on the sides to assist identification and to enable the face of the tally to be distinguished from its back. Both face and back were used for the financial records and both pieces recorded identical information. Menninger describes how an account was typically settled:

> If the official in charge of a district owed the Crown £100 annually in taxes and duties, he would make the first payment at Easter, let us say £40. In testimony of this a *tally* for £40 would be carved, of which the official retained the *stock* (or *stipe*) as a receipt for his payment and the court of audit kept the inset piece (the *foil*) for its own records. The name of the payer and the nature and amount of the payment were incised on both pieces. Then on Michaelmas-day (September 29) the official had to make good the entire annual levy of £100. To do this, he would submit his *stock* for the amount of £40 and pay the rest. The stock was very carefully compared with the *foil* in the Treasury's possession; if they agreed, the amount was written up to the payer's account. If any deception or cheating was detected, the official would be arrested on the spot. The payments made by the county officials were entered in the receipt book, and when their obligations were discharged their accounts for the year were closed with the phrase: *et quietus est* or *and he is quit* (of his debt).[27]

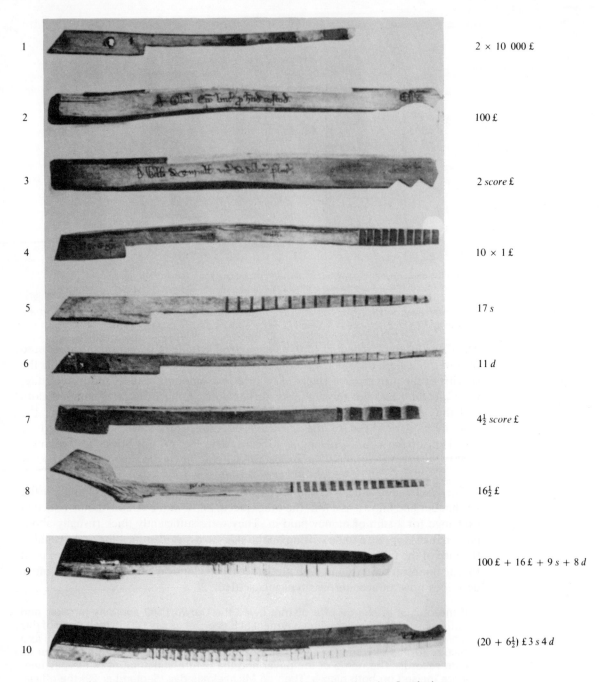

FIGURE 26   British Exchequer tallies (Society of Antiquaries, London)

The fact that the foil was retained made forgeries difficult but they did occasionally occur.

> One instance of forgery, however, occurred in 1297 when William de Boochose was entrusted by the sheriffs with sixty shillings in cash and a tally recording five marks (an unspecified amount which had already been paid to the Exchequer). He pocketed the cash and added the equivalent notches for sixty shillings to the tally; but his dishonesty was discovered and he was sentenced to prison for a year and a day.[28]

Tallies very soon came to be used much as bank notes or bearer cheques are used today and they passed from one person to another in the settlement of private debts. From around 1300 the Exchequer used them as a means of settling with its own debtors and creditors simultaneously by compelling the debtor to pay the creditor direct.

There is an interesting record of William Trente, a steward of Edward I (*r.* 1272–1307), who was in need of a sum of money. He was given a tally representing part of a large sum of money owed to the Exchequer by the citizens of London. The steward was then to demand the amount represented by the tally from the citizens before handing over the tally itself and the citizens were notified accordingly.

This cashless system of exchange was in constant use from the middle of the fourteenth century and it declined only with the rise of the banking institutions in the eighteenth century.

The carving of notches in the tallies was uniformly regulated. We are well informed about this by the *Dialogus de Scaccario* ('Dialogue concerning the Chessboard') written in 1186 by Richard, Royal Treasurer and Bishop of London. In the chapter 'On the cutting of the tallies' the *Dialogue* states:

> The notch for £1000 is placed at the end and is as large as the hand is wide (no. 1 in illustration, p. 46);
> for £100 the notch is as large as the thickness of a thumb, and to distinguish it from that for £1000 it is not straight but curved (no. 2);
> for £20 it is as large as the thickness of the little finger (no. 3);
> for £1 it has the breadth of a ripe barleycorn (no. 4);
> for 1 shilling it is smaller but still large enough to be seen as a notch (no. 5);
> for 1 penny only a cut is made, with no wood being removed (no. 6);
> for a half of any of thse units a notch or cut half the length is carved: one cut slantwise, one perpendicular to the edge (nos 7, 8)

The highest numbers were carved into the bottom surface (like the notch for £100 in no. 9) but the smaller ones at the top from left to right so that the lowest ranks stand above the highest, from which they are clearly differentiated.

On the tallies it can be seen that the intermediate rank between £100 and £1 was not £10, as might be expected, but £20. Thus, appears the old 20-grouping again in the *scores*: a *score of pounds* is £20. But what is a *score*? As shown earlier, the English word *score* is derived from old Saxon *sceran*, to cut. Thus, a *score* is something which has been cut or carved: a notch to be incised on a tally stick whenever a group of 20 was counted. From this notch the group of 20 received its English name *score*.

To run into scores means 'to go into debt', and *on score* means 'on credit'. There are many passages in Shakespeare in which the word *score* still has its original meaning 'notch':

> Our forefathers had no other books than the score and the tally. (*Henry VI Part 2,* Act IV, Scene 7).
>
> I thank you, good people. There shall be no money, all shall eat and drink on my score. (Act IV, Scene 2)
>
> He parted well and paid his score. (*Macbeth,* Act V, Scene 7)

The wooden Exchequer tallies fell into disfavour with the advent of paper money. In 1694 the estimated total value of the tallies was £15 000. In 1697 the Bank of England increased its capital and a large number of the shares were purchased with wood tallies. C. R. Josset writes:

> Some of the tallies represented payments to government departments, including some abnormally large ones. One of these, retained at the Bank of England Museum, is eight feet six inches long, and represents the maximum amount to which one tally was restricted – £50,000. It owes its existence to the fact that it had never been returned to the Exchequer, as the payment represented a government debt which had never been paid![29]

The use of tallies was abolished by statute in 1783 but the statute was not strictly enforced and tallies continued in use until 1826. In 1834 much of the evidence of the Exchequer tallies was lost when a very large number of old tallies were burned in the furnaces which heated the House of Lords. These tallies ended their history with some notoriety since the fire kindled with them started a more general conflagration which resulted in the destruction of the old Houses of Parliament.

On the Finnish tally stick shown in Figure 22 simple strokes and crosses can be seen. We may supposed that the single strokes mean 1 and the crosses 10 like the Roman numerals I and X.

On the Swiss tally sticks shown in Figure 21 there is another type of crossed stroke. On the second stick there are five single strokes crossed by a long line, and the next two single strokes. The value indicated is

$$50 + 2 = 52$$

A bundle of Alpine 'number billets' is shown in Figure 27. On the second and third of these, there are what appear to be the Roman numerals:

$$XXXVIIII = 39$$
$$XXXVII \quad = 37$$

On the last and most ornate tally stick the number 122 is written in the following way: the number VVI, which means $5 + 5 + 1 = 11$, is crossed by a long stroke which means that its value is to be multiplied by 10 (as with the tally stick shown in Figure 21 where five single strokes crossed by a long bar means 50); next, XII is added to give

$$110 + 12 = 122$$

On the other sticks shown halved crosses and halved single strokes occur indicating

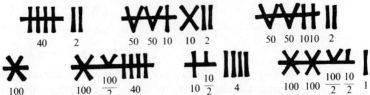

FIGURE 27 Bundle of Alpine number billets, small flat sticks some 20 cm long on which are carved the cow rights to which the owner is entitled; the owner's name or symbol is on the reverse side. The most ornate stick at the extreme right shows the total number, 122. This bundle, from Saanen in the canton of Berne, is dated 1778 (by courtesy of Vandenhoek & Ruprecht, Göttingen). Underneath is an illustration of the whole system

that the values 100 and 10 are to be halved. The whole system is illustrated in the lower part of Figure 27.

It is interesting to note that in China a similar method was used. On wooden sticks from the time of the Han Dynasty (200 BC to AD 200) the numbers 10, 20, 30 and 40 are marked in just the same way as on the Swiss tally stick (see Figure 28).

FIGURE 28 The ten stroke on Chinese Han sticks

The people who used the Swiss tally sticks shown on pages 43–5 could in fact write, for there are the letters HW on the left-hand stick in the bundle of Alpine number billets (see Figure 27). There is no doubt whoever invented the system knew and used the Roman numerals I, V and X. The sign C for 100 was not so easy to carve and so the method of crossing the signs I, V and X was invented.

So far everything seems clear. But now the question arises: how can one explain the Roman numerals themselves? Were they derived from the Roman letters I, V, X, L and C or were they derived from notches on tally sticks?

The numerals C and M may be explained as the first letters of *centum* and *mille* but I, V and X are certainly not the first letters of *unus, quinque* and *decem*. On the other hand in carving notches it is quite natural to use a single stroke 1 for *one*, two crossed strokes × for *ten* and a halved cross V for *five*.

The same method of halving symbols was also used in the case of 1000 and 500. An old engraved symbol for 1000 was C|Ɔ , which was later simplified to M. The right half of this symbol was |⊃ , whence the Roman numeral D for 500.

The origin of the sign L for 50 is also uncertain. An inscription on a Roman milestone reads:

> I built the road from Regium to Capua and erected all the bridges, milestones, and postal stations on it. From here to Novceria the distance is 51 miles, to Capua 84, to Muranum 74, to Consentia 123, to Valentia 180, to the statue on the seashore 237 and from Capua to Regium in all 321 miles. Likewise as Praetor in Sicily I tracked down Italic fugitive slaves and returned 917 persons (to their masters). Moreover I was the first to establish that on public lands herdsmen must yield to farmers. At this place I built the forum and other public buildings.[30]

Number 51 is written ↓1, 74 as ↓XXIIII and 180 as C↓XXX. A possible derivation for the L (= 50) is:

$$↓ \longrightarrow \Downarrow \longrightarrow \underline{↓} \longrightarrow L$$

Roman numerals are discussed in detail in Chapter 4.

Carving notches in tally sticks was perhaps the earliest form of recording numbers and, as mentioned earlier, Menninger confidently states that carvings on a notched stick are 'folk-signs, number symbols devised by simple people themselves to meet their everyday needs', but is this opinion justified in all cases?

It is true that specific kinds of tally sticks were used only in limited areas. However, the tally sticks from the canton Berne (see Figure 27) have Roman numerals and so do the tally sticks from Visperterminen, shown in Figure 25. On the 'milk sticks' from Graubünden (see Figure 21) as well as on the tally sticks from Berne capital letters are carved and on the 'Alpine billet' (see Figure 24) is the date 1752 written in Arabic numerals. So the inventors and users of these tally sticks were not just 'simple people'; at least some of them could read and write.

By examining the words used to designate tally sticks in Europe one can safely conclude that the English, French, Spanish and Italian tally sticks go back to ancient Roman times at least, and that tally sticks were known in ancient Germania before the North and West Germanic languages separated. It has already been said that the English word *tally* comes from the Latin *talea* through the French *taille*. The Spanish and Italian words, *tarja* and *taglia*, are also derived from the Latin *talea*. The Swedish word *karvstock* is the same as Dutch *kerfstok*, and the German word *Kerbholz* is another variant. This means that the North and West Germanic language families had a common word for 'tally stick'.

All this points to the conclusion that tally sticks were not independent local inventions. They have a long tradition and it is even possible that in the region of the Indo-European languages they were invented only once just as the decimal counting system was invented only once within the same region.

Knots

Another widespread method of representing numbers involves the use of *knots*. However, unlike the tally marks, such knots have never evolved in any way into written symbols for numbers. Two present-day examples of forms of 'number strings' are the Rosary and the Tibetan prayer strings by means of which religious exercises are counted.

In the *History*, IV, Herodotus records that the Persian king, Darius, gave the Ionian princes a cord with 60 knots tied in it when he set off to conquer ancient Greece. They were to guard the bridge of ships over the River Istros, untying one knot each day until he should return. If he had not returned when the last knot was untied they were to await him no longer.

A series of identical knots tied on a single cord is exactly equivalent to the carving of unit notches on a tally stick, and needs no further discussion here. There are, however, several forms of number strings which are of particular interest because they go beyond the mere succession of identical knots. Probably the most striking of these are the *quipus* or 'knotted cords' of Peru.

The *quipus* were used in the Inca Empire to record all official land transactions and are therefore similar in purpose to the British Exchequer tallies discussed earlier. They played a crucial role in the Inca Empire because no system of written numerals was known there.

Three different kinds of knot were tied: single knots, double knots and slip knots with from 2 to 9 loops. The knot system was decimal in that the knots were arranged in groups, with those representing hundreds nearest to the main strand at the top, tens in the middle and units at the bottom. The headstring was threaded through loops at the top of the strings and its end was knotted so as to give the sum of the various numbers represented. This is shown in Figure 29.

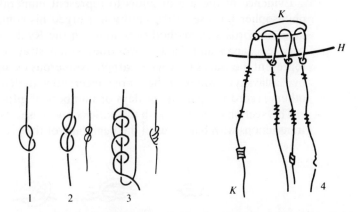

FIGURE 29   The various knots in a Peruvian *quipu*: 1   single or overhand knot; 2   double or figure-of-eight knot; 3   slip-knot with three loops; 4   the head string marked K, which is run through the loops of the other strings showing the numbers 150, 42 and 231, bears their sum, 423

Menninger comments on the mechanics of the offical *quipus* records as follows:

In every Inca settlement there were four official *quipu* keepers, known as *camayocs*, who tied the knots in these strings and submitted them to the central government in Cuzco. There is no doubt that this perhaps intentionally obscure manner of recording numbers, which only initiates could read, was a strong support for the monarchial absolutism of the Inca ruler. An example of the opposite situation is provided by ancient Athens, where the government of the city-state was obliged to reveal its records openly to the citizens so that every official was subject to 'democratic' criticism − dictatorship and democracy in book keeping.[31]

It is tempting to wonder if the Inca *camayocs* were as unpopular as the publicans (i.e. tax gatherers) of the time of Christ!

The Incas also recorded non-numerical data on their *quipus*, though this could lead to considerable misunderstanding. Garcilaso de la Vaga, whose father was Spanish and mother an Inca princess, records the reception of the Spanish ambassador, Hernando de Soto, by the Inca, and the subsequent translation of his message:

> Among the common people and the nobility who attended the Inca in the audience hall there were two official historians who recorded Hernando de Soto's message and the Inca's reply in knots.
>
> .   .   .   .
>
> His translation was not good and accurate; this was unintentional, of course, since he did not understand the meaning of what he had to translate. Instead of the three-in-one and single God he read three gods plus one makes four (!) adding the numbers so as to make the expression intelligible to himself.[32]

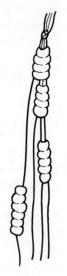

FIGURE 30   *Chimpu* showing the number 4456

The *chimpu* of the Bolivian and Peruvian Indians derives from the Inca *quipu*. Numbers are recorded using fruit seeds strung like beads. The cords are first tied at the top and the seeds representing thousands are strung on four cords, those representing hundreds on three, those representing tens on two and those representing units on one (see Figure 30).

Evidence of the use of knots to represent numerals is universal. The Chinese philosopher Lao-tse (fifth century BC) urged his compatriots to go back to tying knots in cords as a method of writing. In the Ryuku Islands in the Pacific Ocean between Japan and Taiwan workmen braid straw or reeds in various ways to represent the wages they have earned. Numerous examples from Africa are quoted by Zaslavsky.[33] Many of these are concerned with debts of one sort or another but others record the passage of days or numbers of gifts.

This section concludes with the *miller's knots* used in Germany by millers in their transactions with bakers until the beginning of the present century (see Figure 31).

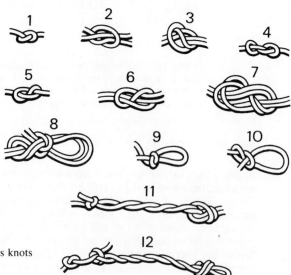

FIGURE 31   Miller's knots from Baden

Menninger describes these as follows:

> The miller had to 'write down' somehow the amounts and the kind of flour and meal contained in the sacks he delivered. For this purpose he would use the draw strings that tied the mouths of the sacks. The quantities and measures were indicated by knots (1–7 in the figure) and the kinds of meal or flour by loops or tufts (8–12). The miller's measure of flour was the *Sester*[34], an earlier measure of volume containing 10 *Mässel*.
>
> | | | | |
> |---|---|---|---|
> | 1 *Mässel* | = | one simple overhand knot | (1) |
> | 2 *Mässel* | = | the same knot with a strand drawn through it or tied in the bight of the draw string. | (3) |
> | 5 *Mässel* | = | $\frac{1}{2}$ *Sester*, and | |
> | 10 *Mässel* | = | 1 *Sester* both represented by special knots | (4, 5) |
> | 2 *Sester* | = | the same 1-*Sester* knot tied in the bight or with a strand drawn through it | (6) |
> | 6 *Sester* | = | likewise represented by a special knot | (7) |
>
> Here for the first time we see that different numbers can be represented not only by series of the same knot, but also by special individual knot-symbols, which thus become knot-numerals: 1, 2, 5, 10, 20, and 60 measures. The intermediate numbers are then generally formed by combinations: $8\frac{1}{2}$ *Sester* = $6 + 2 + \frac{1}{2}$ *Sester*.
>
> The manner of indicating the kinds of meal and flour delivered by the miller, for instance 'hog-mash, rye, barley, 1st and 2nd grade wheat flour' can be seen from the illustration (8-12). Barley rye, for example, was represented by a loop tied into a *Sester*-knot.[35]

As a postscript to the earlier discussion of tallies, there is an interesting and unusual case which may be noted where the principle of tallying was adapted to the representation of words. This occurred amongst the Celtic races, especially in Ireland and Wales, in the early days of Christianity, and involved cutting lines from the edges of large stones to represent letters of the alphabet. In this *Ogham script*, as it is known, the lines were cut from the edges either orthogonally or obliquely as shown in Figure 32.

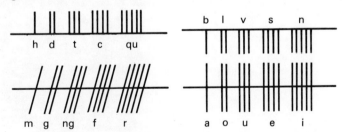

FIGURE 32 The Ogham alphabet (the horizontal lines represent the edge of the stone)

From one to five cuts were made, either into one of the sides from the edge or into both sides, representing twenty letters in all. The direct link with numerical tallying is provided by the top-left group. Here, the representations of *h, d, t, c* and *qu* correspond to the initial letters of the earliest known form of the number words: *a hoina* (1), *a duou* (2), *a ttri* (3), *a ccetuor* (4), and *a qqueqque* (5). However, this principle could not be extended further, so the remaining letters were grouped in fives: the soft consonants (top right) and the vowels (bottom right), the rest having the angled cuts (bottom left).

Variations of this code can be found. For example, the vowels are sometimes just very short cuts across the edges as in the case of the Irish gravestone in Figure 33, which reads (from bottom right around the edges to bottom left) *d o d i k [a] m a qu i m a qu*. The bottom part of the stone has been broken off. The complete text could have translated as 'a [grave of --] dodik, son of maqu [---]'. Note the crossed incision for *k*, a late form.

FIGURE 33   Irish gravestone from Aglish with Ogham characters (National Museum of Ireland)

Such gravestones have quite often been removed from their original sites and used as sills of church windows or, in some cases, simply built into the walls of churches. The earliest examples of these stones date from before the fourth century and this method of writing died out some six centuries later. The key for deciphering Ogham texts can be found in a fourteenth-century vellum manuscript, *The Book of Ballymote*, but, despite this key, scholars have often encountered difficulties in determining the precise meanings of the texts. Apart from occasional notes in manuscripts, no literature in Ogham script exists. References to Ogham in various manuscripts indicate that it was used for writing messages on wooden staves.

The origin of counting

So far the *origin of counting* as such has not been discussed. There are good reasons for this. The validity (or otherwise) of theories of the origin of counting is dependent to a considerable extent upon a satisfactory outcome of the *independent invention* versus *diffusion* controversy. Although this chapter presents considerable evidence which, in the main, can be held to support the diffusion hypothesis, there is no suggestion that the controversy is entirely resolved.

The *independent inventionist* will look for the origin of counting in human need to come to terms with and conquer the environment – that is, in practical applications of number. The *diffusionist*, on the other hand, seeks the origin in a single centre of civilisation, and is therefore by no means restricted to a purely practical basis for the invention of counting.

There has been no attempt to present the various *origin* arguments with their supporting evidence because the independent inventionists, for the most part, appear to take their hypothesis for granted and then seek for highly speculative psychological explanations of it. These explanations purport to determine what went on in the minds of our forebears many thousands of years ago and are not supported by any historical facts.

Seidenberg has published a second paper, which is directly concerned with the origin hypotheses. The argument concludes with the following sentence:

> Counting was invented in a civilised center, in elaboration of the Creation ritual, as a means of calling participants in ritual onto the ritual scene, once and only once, and then diffused.[36]

The arguments supporting this hypothesis are interesting but not conclusive (as Seidenberg himself admits) since much of the evidence considered is also compatible with a practical origin of counting. The paper itself should be referred to if the reader is interested in a detailed study of these arguments.

3 Number Words

In Chapter 2 the discussion of the counting systems of primitive tribes was confined largely to what is known of their present or recent state. The history of these counting systems is not known in any detail. Hypotheses about the *distribution* of these systems, interesting though they are, cannot truly be classified as 'history'. They are speculative and uncertain. However, for the Sumerian and Babylonian number system, there are *written* sources going back to around 3000 BC. The gradual development of this system from 3000 BC until around 1700 BC (the time of Hammurabi[1]) can therefore be studied. After 1700 BC the Babylonian system remained the same until the time of Christ.

Linguistic principles

Turning to the decimal counting system, the situation is still more favourable because one can draw on the science of *linguistics*. English belongs to the family of *Indo-European* languages, the development of which can be followed over a period of four or five thousand years. All Indo-European languages have essentially the same counting system and they all stem from one mother language, or at least from one group of closely related dialects, which already existed between 3000 and 2500 BC. Therefore, in order to study the history of the decimal counting system, one must study the history of the Indo-European languages.

The first question that arises is: How can kinship between languages be recognised? One can start with two familiar examples of language groups, the *Romance* and the *Germanic* languages, and compare the various number words within these groups.

In comparing languages it is very useful to compare the number words first. Other words often change their meaning or go out of use. For instance, the Dutch word *zee* means sea but in High German the same word *See* is used to denote a lake. Such changes of meaning very seldom happen with number words. Hence, number words are good objects to consider in comparing languages.

Table 2 shows the number words for 1, 2, 3, 4, 5 in four modern Romance languages.

TABLE 2   Number words for 1–5 in four modern languages

	Italian	French	Spanish	Romanian
1	*uno, una*	*un, une*	*uno, una*	*uno*
2	*due*	*deux*	*dos*	*doi, doua*
3	*tre*	*trois*	*tres*	*trei*
4	*quattro*	*quatre*	*cuatro*	*patru*
5	*cinque*	*cinq*	*cinco*	*cinci*

Even a superficial glance reveals the likeness of these number words. The explanation for the likeness is well known. All the Romance languages are derived from Latin. The Latin sequence is:

1 *unus, una, unum*
2 *duo, duae, duo*
3 *tres, tria*
4 *quattuor*
5 *quinque*

The Roman soldiers spoke a simplified Latin called 'Vulgar Latin'. They were sent out to all parts of the Roman Empire: to Spain, France and to the whole region between France and the Black Sea; traders and settlers also came to these provinces. Thus, vernacular Latin spread from Italy to Spain, France, Raetia[2] and Romania.

Even if these historical facts were not known, one might conclude from the similarity of the number words (and of the other words of these languages) that all Romance languages were akin.

The comparison also shows how languages are simplified in the course of time. The neuter forms *unum, duo* and *tria* have gone out of use. The Latin ending *-us* becomes *-o* in Italian, Spanish and Romanian and disappears in French. The *s* at the end of *tres* has disappeared in Italian and Romanian, and it is no longer pronounced in French.

The oldest known Germanic language is Gothic. It is the language used by Bishop Wulfila (or Ulfilas) in his translation of the Bible written about AD 350. It is very instructive to compare the High German number words from 2 to 10 with the Gothic ones, leaving out the number words for 1 because of complications of inflection (Table 3).

TABLE 3 High German and Gothic words from 2 to 10

	Gothic	High German		
		Old	Middle	New
2	*twai, twos, twa*	*zwene, zwa, zwei*	*zwene, zwo, zwei*	*zwei*
3	*threis, thria*	*dri, drio, driu*	*dri, driu*	*drei*
4	*fidwör*	*fior*	*vier*	*vier*
5	*fimf*	*fimf or finf*	*fünf*	*fünf*
6	*saihs*	*sehs*	*sehs*	*sechs*
7	*sibun*	*sibun*	*siben*	*sieben*
8	*ahtau*	*ahto*	*ahte*	*acht*
9	*niun*	*niun*	*niun*	*neun*
10	*taihun*	*zehan*	*zehen*	*zehn*

The likeness between Gothic and Old High German is so great that a common origin cannot be doubted. From the Old High German forms the later ones were derived by gradual simplification; this is the same phenomenon observed in the Romance languages.

Adjectives and the dual

At a very early stage of counting, the 'number' of something that was counted was felt to be one of its attributes. Thus number words could be expected to appear as adjectives.

In Gothic as well as in High and Middle German the number words for 2 and 3 do appear as adjectives; they have separate masculine, feminine and neuter forms, and they are inflected[3] like adjectives. The same is true for Latin and Greek. Table 4 below shows the forms of *three* in Greek, Latin and Gothic.

TABLE 4   Forms of *three* in Greek, Latin and Gothic

Case[4]	Greek	Latin	Gothic
Nominative	*treis, tria*	*tres, tria*	*threis, thrija*
Accusative	*treis, tria*	*tres, tria*	*threis, thrija*
Genitive	*triôn*	*trium*	*thrije*
Dative	*trisi*	*tribus*	*thrium*

Table 3 listed the various forms of *two*, according to gender, in Gothic and in Old and Middle High German. In some German dialects the distinction between the three forms of *two* can be observed even today. For example, in Upper Hesse people say:

> *zwien Osse* = two oxen
> *zwoo Käu* = two cows
> *zwaa Kinner* = two children

In Martin Luther's translation of the Bible there are the forms:

> masculine *zween*
> feminine *zwo*
> neuter   *zwei*

for example:

> *Niemand kann zween Herren dienen*
> 'No man can serve two masters' (Matthew 6:24)

and

> *Und stand auf in der Nacht und nahm seine zwei Weiber und seine zwo Mädge*
> 'And he rose up that night, and took his two wives, and his two women servants' (Genesis 32:22)

In Greek the number word for 4, for example, has both gender changes and inflection:

Nominative	*tettares*	*tettara*
Accusative	*tettares*	*tettara*
Genitive	*tettarôn*	
Dative	*tettarsi*	

Gender changes and inflection provide examples where number words are still intimately bound up with the objects counted. Such intimate connection is a sure sign of great antiquity. To illustrate this consider the relation between adjectives and nouns in English, French and Latin. In English, which is a modern, highly simplified language, one says:

> *a good father*
> *a good mother*

The word *good* is always the same: it does not indicate the gender of the noun. In French the masculine and feminine forms are different:

> *un bon père*
> *une bonne mère*

but the forms *bon* and *bonne* are the same in nominative, accusative, genitive and dative. In Latin, gender influences the form of the adjective:

> *pater bonus*
> *mater bona*

and in addition to this the adjective changes its form in the other cases (accusative, genitive, dative, etc.) and also in the plural. This change of form, which is found in all ancient languages of the Indo-European family, means a close connection between adjective and noun.

The same phenomenon can be observed in the case of number words. The number word for 4 is inflected like an adjective in Greek, but not in Latin and Gothic. Now Greek is a relatively ancient language: it was spoken (and even written) as early as 1400 BC on the island of Crete, Thus the inflection of number words is a sure sign of antiquity: it fell out of use in later times.

In several ancient languages not only a singular and a plural are to be found, but also a *dual*, which was used in the case of just two objects. A very ancient, though rare, dual can be found in classical Greek:

> *ho philos*   'the friend'
> *to philô*   'both friends'
> *hoi philoi*   'the friends'

Ancient Semitic languages also had a distinct dual (found, for example, in the Hebrew Bible).

A remnant of an old dual is found in the Russian language:

> *dva doma*   'two of house'
> *tri doma*   'three of house'
> *četýrě doma*   'four of house'

but

> *pjatj domov*   'five of houses'

With numbers beyond 4 the genitive plural *domov* is used which is perfectly logical. The form *doma* used with 2, 3 and 4 resembles a genitive singular but it is really an old dual form.

The use of the dual is logically correct when the number of objects is 2. However, in popular speech the dual form was extended to 3 or even 4. The use of the dual also indicates a close relation between the number itself and the objects being counted. In this case it is the number which determines the grammatical form of the noun.

In all known Indo-European languages, numbers beyond 4 are not treated as adjectives: they remain completely unchanged. How can one explain the break after 4? This question is discussed by Menninger as follows:

> Why do only the first four numbers, 'one', 'two', 'three', and 'four', appear as
> inflected adjectives? Why not 'five' or 'seven' or 'twenty'?
>   We can very well answer this ourselves: Because they were the earliest number
> words (not counting the very first step forward, from One to Two, which had still
> taken place completely within the realm of the mind).
>
> . . . .
>
> One may then ask: Of course this is natural enough, but why should the break come
> just after Four? Why not after Seven? Two reasons may be given. The first is that
> the hand has four figures, not counting the thumb. What happened to the thumb
> here was like what happened to One – it was not regarded as being equal, it was not
> a 'finger' like the others. The handsbreadth, measured without the thumb across
> the knuckles, was used as a basic measure by almost all ancient civilisations. The
> Greek and Egyptian *ell*, for instance, has 6 handsbreadths, or $6 \times 4 = 24$ fingers;
> likewise Roman foot (*pes*) was made up of 4 *palmae* and of 4×4 *digiti*. A second
> reason might be that a quantity larger than four, or even three, can no longer be
> directly apprehended. If we ask, 'How many people were there?', the answer is
> 'three or four', not 'nine or ten', for that would already be a 'large number'. And in
> those early primitive times only clearly perceptible numbers were apprehended as
> words, as we have seen in the case of the dual, which expresses a distinct Two, the
> 'other' that goes with the One. Four, then, was certainly another limit of counting.[5]

Consider the meaning of the expression 'limit of counting' a little more closely. If
there are independent number words up to ten, composite number words like
'nineteen' can always be formed. The Romans had number words up to *mille* =
1000, but they had no difficulty in forming composite number words like *decem
milia*, which literally means 'ten thousands'. It follows that a 'limit of counting' is
never an absolute limit: it can always be transcended by forming composite number
words.

In this sense Menninger may well be right. It is quite possible that in the original
Indo-European language 'one', 'two', 'three' and 'four' were earlier number words,
which were more closely connected with the counted objects and treated like
adjectives, whereas 'five', 'six', etc. were later number words, perhaps taken over
from a foreign language, which were no longer considered as adjectives and hence
remained unchanged in all cases.

Another argument in favour of Menninger's hypothesis is given by the form of
the number words for 8 in Latin and Greek. In Latin the word is *octo*, in Greek
oktô; the ending -*ô* in Greek is a dual ending. In the Gothic and Sanskrit words for
8, *ahtaú* and *aštau*, the ending -*au* is also a dual ending. So it seems that 8 was
originally conceived as a dual, i.e. as two fours, although the first syllable in *okto*
cannot be recognised as meaning 'four'.

In many primitive languages, the number word for 8 is formed as 'twice 4'. So it is
quite possible that in the mother language of the Indo-European family the
independent number words originally ended at 4 and that 8 was expressed as
twice 4.

It is also possible that at one time counting stopped at 8. This would occur
naturally when two handsbreadths had been used up in measurement. After the
doubling of 'four' to give 'eight', there would be a need for a 'new' number before
counting could continue. It is a striking fact that there is a similarity in most
Indo-European languages between the word for 9 and the word for 'new'. Thus we

have for example:

'nine': Sanskrit *nava,* Latin *novem,* Gothic *niun,* Tokharian *nu*
'new': Sanskrit *navas,* Latin *novus,* Gothic *niujis,* Tokharian *nu*

Such evidence is of course not absolutely conclusive.

Reconstruction of original Indo-European words

In one respect High German differs from Gothic and from all other Germanic languages. The *t* occurring in 'two' and 'ten' (Gothic *twai* and *taihun*) is replaced by a *ts* (written as *z*) in High German. Thus, Gothic *twai* becomes *zwei,* and Gothic *taihun* becomes *zehn.*

This phenomenon is called the *Middle German phonetic shift.* It is characteristic for High German; it did not take place in Dutch, nor in the Low German dialects of Northern Germany:

English *ten,* Dutch *tien,* German *zehn*
English *water,* Dutch *water,* German *Wasser*
English *pound,* Dutch *pond,* German *Pfund*
English *make,* Dutch *maken,* German *machen*

The Middle German phonetic shift is not a shift according to a stringent rule. It is better to describe it as a tendency, prevalent in Southern Germany and Switzerland, to replace

t by *ts* (*z*) or *s*
p by *pf* or *f*
k by *kch* or *ch*

It is difficult to date this phonetic shift with any real accuracy but the differentiation between High and Low German was probably definitive by the beginning of the seventh century. The line of differentiation runs through the centre of Germany from Aachen to Frankfurt-on-the-Oder, roughly through Düsseldorf, Kassel and Magdeburg (see Figure 34). It crosses the Rhine at Benrath and is therefore sometimes called the *Benrath line.*

The Middle German phonetic shift sometimes makes it difficult to recognise the

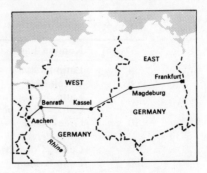

FIGURE 34   Central Germany showing the area of the 'Benrath line'

similarity between High German and other Germanic languages. The comparison which follows therefore leaves out of account the High German number words and is restricted to languages such as Gothic, Dutch, Danish and English which were not affected by this shift.

In English many words are derived from French or Latin but the number words are Germanic. This becomes clear if we compare the English number words with those of other Germanic languages. Table 5 includes only the masculine forms of 2 and 3.

TABLE 5   Number words of various Germanic languages

	East Germanic	North Germanic				West Germanic	
	Gothic	Icelandic	Old Nordic	Danish	Swedish	Dutch	English
2	twai	tveir	tveir	to	tva	twee	two
3	threis	thrir	thrir	tre	tre	drie	three
4	fidwor	fjorir	fjorer	fire	fyra	vier	four
5	fimf	fimm	fimm	fem	fem	vijf	five
6	saihs	sex	sex	seks	sex	zes	six
7	sibun	sjö	siau	syv	sju	zeven	seven
8	ahtau	atta	atta	otte	åtta	acht	eight
9	niun	niu	nio	ni	nio	negen	nine
10	taihun	tiu	tio	ti	tio	tien	ten

It is quite clear that all these languages belong to one and the same language group. Icelandic is very similar to Old Nordic; it has preserved many archaic features. Danish and Swedish have more modern, simplified forms. Dutch is very similar to Low German (*Plattdeutsch*). The English number words are in many respects similar to the Dutch ones but they also have some features in common with Old Nordic such as the *th* in *three* and the *o* in *four*. In the Anglo-Saxon language, as it was spoken in England in the twelfth century, the similarities with Old Nordic were even greater. Thus, *four* was in Anglo-Saxon *feower*, in Old Saxon *fiuwar* or *fior* and in Old Nordic *fjorer*.

All Indo-European languages can be classified as *Kentum* or *Satem* languages. The Latin word *centum* (pronounced as 'kentum') means 'hundred' and *satem* is the corresponding Old Persian word. Table 6 compares some number words from

TABLE 6   Number words from Kentum and Satem languages

	Kentum Languages	Satem Languages
4	Latin *quattuor*	Sanskrit *čatvarah*
8	Greek *okto*	Sanskrit *aštau*
	Latin *octo*	Lithuanian *aštuoni*
	Celtic (Irish) *ocht*	
10	Greek *deka*	Sanskrit *daśa*
	Latin *decem*	Lithuanian *dešimt*
	Celtic (Cornish) *dek*	Czech *deset*
100	Latin *centum*	Sanskrit *šatem*
	Celtic (Breton) *kant*	Lithuanian *šimtas*
	Tokharian *kânt*	

Kentum and Satem languages. The characteristic consonantal differences between
Kentum and Satem words for 4, 8, 10 and 100 show that where the Kentum
languages have *k* or *ch* (hard) or *qu* (= *kw*), the Satem languages have *s* or *š* or *č*
(soft *ch*). The shift from *k, ch, qu,* to *s, š, č* is characteristic for many number
words.

In the list below, Kentum and Satem languages are collected in groups of closely
related languages. *Dead languages* are marked with an asterisk; the oldest known
of these is Hittite, the language of the Hittites, who are supposed to have entered
Cappadocia (in Anatolia, Central Turkey) about 1800 BC.

Kentum languages

Anatolian (Hittite*, Luwian*, etc.)
Greek (Old* and new Greek)
Italic (Latin*, Italian, Spanish, Portuguese, French, Rhaeto-Romanic, Romanian)
Celtic (Gaulish*, Breton, Cornish*, Welsh, Manx, Irish and Scottish Gaelic)
Germanic (Gothic*, Old Norse*, Icelandic, Norwegian, Swedish, Danish, Frisian,
Anglo-Saxon*, Old Saxon*, English, Dutch, Low German, High German)
*Tokharian** (Spoken in Western China until the seventh century AD, known from
manuscripts found in Turkestan)

Satem languages

Indic (Vedic*, Sanskrit*, many languages spoken in India)
Iranian (Avestan*, Old Persian*, Middle Persian*, Persian, Afghan, Kurdish, etc.)
Slavic (Old Church Slavonic*, Russian, Ukrainian, Polish, Czech, Slovak, Serbo-
Croatian, Slovenian, Bulgarian, etc.)
Baltic (Old Prussian*, Lithuanian, Lettish)
Thrako-Illyrian (Illyrian*, Thrakian*, Albanian)
Thrako-Phrygian (Phrygian*, Armenian)

It is not known in what part of Europe or Asia the mother language of all
Indo-European languages was spoken. What is known is that between 2500 and
2000 BC a population, speaking Indo-European languages, was firmly established in
a region extending from Middle Germany (Thuringia) to Southern Russia. On the
map in Figure 35 this region is roughly indicated by two rectangles. In the Western
part of this region Kentum languages were spoken, in the Eastern part Satem
languages. From this central region the Kentum and Satem languages spread out
over Europe and parts of Asia.

It is not known when these migrations began or what track they followed,
therefore the details on the map are not to be taken too seriously! Only the
migrations in north-west Europe, which took place in historical times, can be traced
with a reasonable degree of certainty. Solid lines indicate the spread of Kentum
languages, hollow lines that of Satem languages. The lines point towards regions
where the languages in question were actually spoken in historical times.

The first known migration was that of the *Greeks*. They arrived in Greece in
several waves between 2400 and 2000 BC. Another early migration, that of the

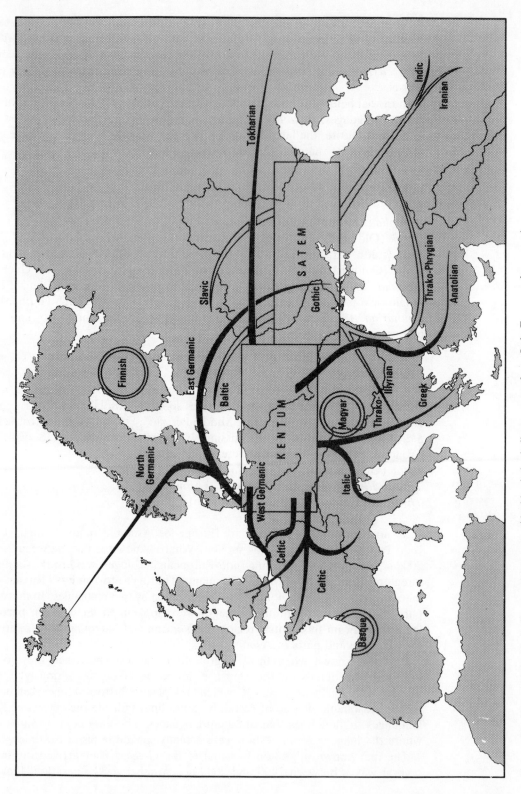

FIGURE 35   The spread of Indo-European languages (double circles indicate areas of non-Indo-European languages)

Hittites, took place between 2000 and 1800 BC. The Hittites settled in Anatolia and founded an empire.

The *Indic* and *Iranian* languages are closely related. The people who spoke these languages were called Aryans in India as well as in Iran. The ancient name of Iran was Ariana, which means 'country of the Aryans'. One group of Aryans came to India about 1500 BC; another group settled in Iran and founded the Persian Empire about 530 BC. Ancient Indic languages were Vedic and Sanskrit. Ancient Iranian languages were Avestan (the language of the Avesta, the collection of sacred books of the Zarathustrians) and Old Persian (the language of the inscriptions of the Persian kings).

The ancient *Italic* languages, of which Latin is the best known, have striking similarities with Greek. About 1200 BC several groups of people speaking these languages migrated into Italy, probably from what is now Yugoslavia.

The *Celts* lived in south-west Germany and in the eastern parts of France early in the second millennium BC. They spread rapidly over western and southern Europe, crossing into the British Isles, moving southward to Spain and Italy, fighting the Macedonians in Hellenistic times and finally penetrating into Asia Minor where they became known as Galatians.

Germanic languages are usually divided into three groups: *West, North* and *East Germanic*. To the West Germanic group belong High German, Low German (*Plattdeutsch*, spoken in northern Germany), Dutch, Frisian and English. North Germanic languages are spoken in Scandinavia and Iceland. The best known East Germanic language is Gothic. The Goths came, according to an old tradition, from Gotaland in South Sweden. By the third century AD they were settled in a region north of the Black Sea; in the fourth and fifth centuries the West Goths (or Visigoths) went farther west and finally conquered Spain, while the East Goths (or Ostrogoths) ended up in Italy where they founded a kingdom about AD 500. Bishop Wulfila, who translated the Bible into Gothic, was a Visigoth.

The *Slavic* languages can be divided into three groups:

> *East Slavic* (Russian, Ukrainian, etc.)
> *South Slavic* (Serbo-Croatian, Slovenian, Bulgarian, etc.)
> *West Slavic* (Polish, Czech, etc.)

The *Thrako-Phrygians* probably went through the Balkan Peninsula to Phrygia in Asia Minor and farther east to Armenia. Of the history of the *Thrako-Illyrian* languages little is known; the living language of Albania belongs to this group.

Tables 7 and 8 show the number words up to 'five' in some Kentum and Satem languages. They show clearly that all these languages are related. For instance, the number words for 3 all have the characteristic letter group *tr* (or *thr* or *dr* in

TABLE 7   Kentum languages

	Greek	Latin	Gothic	Breton	Tokharian
1	*heis, mia, hen*	*unus, -a, -um*	*ains*	*eun*	*sas, sam*
2	*dyo*	*duo, -ae, -o*	*twai, twos, twa*	*diou*	*wu, we*
3	*treis, tria*	*tres, tria*	*threis, thrija*	*tri*	*tre, tri*
4	*tettares, -a*	*quattuor*	*fidwor*	*thevar*	*śtwar*
5	*pente*	*quinque*	*fimf*	*themth*	*pän*

TABLE 8   Satem languages

	Sanskrit	Avestan	Old Church Slavonic	Lithuanian
1	ekab, eka	aevo, aeva, oim	jedinu, -a, -o	vienas
2	dvi, dve	dva, duye, dvae	dva, dve	du, dvi
3	trayah, tisrah	thrayo, tishro, thri	trije	trys
4	čatvarah, čtasrah	čathwaro, čatanro, čatura	četyre	keturi
5	panča	panča	pètj	penki

Germanic languages). Where changes occur they can in many cases be explained by vowel or consonantal shifts that are characteristic for the languages in question.

These vowels or consonant shifts can be illustrated by considering the words for 5. They can all be derived from a supposed Original Indo-European word *penque (the asterisk here means 'hypothetical'). The initial p remains p in most languages, but in Germanic it is regularly softened to pf or f, thus:

> Greek and Latin *pater*
> Gothic   *fadar*
> English   *father*

The first vowel in *penque* is an *e*. It usually becomes an *a* in Sanskrit and *qu* regularly becomes *č*, hence the Sanskrit form *panča* is clearly derived from *penque*.

In Germanic the *e* very often becomes an *i*. The *n* in *penque* was retained in Old High German *finf* and Middle High German *fünf* but it was replaced by an *m* in Gothic possibly because *fimf* is easier to pronounce than *finf*. In Dutch and English, the *m* or *n* was dropped: Dutch *vijf*, Anglo-Saxon *fif*, English *five*. In Scandinavian languages the final *f* of Gothic *fimf* was dropped. Thus Old Nordic and Icelandic have *fimm* and Danish and Swedish have *fem*.

In Greek, Indo-European *qu* is usually replaced by *p, t* or *k*. Hence the Greek form *pénte* can be derived from the Original Indo-European form *penque*, as can the Breton and Tokharian forms.

The various comparisons made so far are important for the history of English number words. The kinship between languages of the Romance and Germanic groups and that between the Kentum and Satem families of languages lead to the conclusion that the origin of English number words lies in a basic mother language from which all other Indo-European languages have evolved. This is the language called *Original Indo-European*. The words of this hypothetical language are not supplied by any ancient written text; they have to be reconstructed by applying to known languages the laws of phonetic and linguistic change that were brought to light by comparative linguistics.

Consider the number words for 10 in Table 9. According to the laws of comparative linguistics the *s* in the Satem languages is derived from Indo-European *k*. The *t* in the Germanic languages is regularly derived from Indo-European *d*. Also the first *a* in Vedic and Avestan *dasa* is derived from an Indo-European *e*. The *h* in Gothic (pronounced as *ch* in German *acht* or in Scottish *loch*) is regularly derived from an Indo-European *k*. Hence the first parts of the words for 10 in the different languages, *dek* or *deich* or *teh* or *das* or *des*, can all be derived from a hypothetical Original Indo-European **dek*.

TABLE 9 Number words for 10 in Kentum and Satem languages

	Kentum Languages		Satem Languages	
10	Greek	*deka*	Vedic	*dasa*
	Latin	*decem*	Avestan	*dasa*
	Old Irish	*deich*	Old Bulgarian	*deset*
	Gothic	*taihun*	Lithuanian	*desimt*
	Old Saxon	*tehan*		

What follows after *dek* is in Greek (as well as in Vedic and Avestan) *a*, in Latin *em*, in Gothic *un*, in Old Saxon *an* and in Lithuanian *im(t)*. All these divergent sounds can be derived, by well-confirmed general linguistic rules, from a hypothetical sound *m̥*, a nasal *m*, which had the value of a vowel. Hence the hypothetical Original Indo-European word for 'ten' can be reconstructed as **dekm̥*.

In order to trace the way in which English and other Germanic languages are related to Original Indo-European we need to return again to the concept of *phonetic shift*.

As discussed earlier the *Middle German* phonetic shift is characteristic for the several variants of High German and it occurred in Southern Germany, in the period ending around AD 600. There is also a much older phonetic shift, known as the *Germanic phonetic shift*, which is common to all Germanic languages. This took place sometime between 400 BC and AD 100 and is described by Menninger as

> the furnace in which the Indo-European language was 'melted down' into 'Germanic'.[6]

Its main rules are:

Indo-European *p, t, k*   become *f, th, ch*[7]
  b, d, g   become *p, t, k*
  bh, dh, gh become *b, d, g*

A standard example for the first and third rules is:

Indo-European **bhratar* (hence Latin *frater*)
Gothic   *brothar*
English   *brother*

Examples for the first two rules can be found among the number words in Table 10. They include:

p, t, k → f, th, ch:
*trejes → threis, three
*penque → fimf, five

b, d, g → p, t, k:
*dekm̥ → taihun, ten

The Middle German and Germanic phonetic shifts make it possible to trace English decimal number words back to their roots in (hypothetical) Original Indo-European thus providing important evidence from linguistics about the age of decimal counting. Decimal counting is thus at least as old as the Indo-European family of languages. This linguistic evidence is well documented from around 1500 BC to the present day.

TABLE 10   Examples of the first two rules of phonetic shift among number words

	Indo-European	Gothic	English
3	*trejes	threis	three
4	*quetuor	fidwor	four
5	*penque	fimf	five
8	*oktou	ahtau[8]	eight
10	*dekm	taihun[9]	ten

The origins of 'ten', 'hundred', 'thousand' and 'million'

As shown earlier the words for 10 in Indo-European languages can be derived from the hypothetical Original Indo-European form

$$*dekm = 10$$

Similarly, the words for 100 can be derived from an Original Indo-European form

$$*kmtom = 100$$

In Latin the nasal m would become em, hence the Latin form centum. In Sanskrit and Greek the m becomes a and in Sanskrit the k becomes $ś$ hence the Sanskrit form is śatem. The Greeks combined hen = one with katon = hundred to give

$$he\text{-}katon = \text{'one hundred'}$$

It is interesting to note that the first syllable of *kmtom is identical with the last syllable of *dekm. This identity is preserved in many languages, e.g.

> Greek deka and (he)katon
> Gothic taihun and hund
> Lithuanian dešimt and šimtas

Hence there seems to be a close relation between the number words for 'ten' and 'hundred'. This relation can be examined a little more closely in Gothic, the oldest known Germanic language.

The Gothic words for 70, 80, 90 and 100 are

> 70 sibuntehund (from sibun = 7 and taihun = 10)
> 80 ahtautehund (from ahtau = 8)
> 90 niuntehund (from niun = 9)
> 100 taihuntehund or taihuntaihund

Thus, in the Gothic translation of the Bible by Wulfila, in the parable of the unfaithful steward (Luke 16:7), the Bishop uses the words:

taihuntaihund mitade kaurns   'hundred measures of wheat'

The last part of all these number words is the substantive *taihund* (derived from *taihun* = 10), which means 'decad', a set of ten. The same method of deriving a substantive from a number word by adding a *d* or *t* at the end is also found in other languages. The Germans have a substantive *Geviert*, derived from *vier* = 'four'. The Greeks have a noun *dekas*, genitive *dekados*, meaning a decad, a set of ten objects belonging together. The root *dekad-* of this Greek word is derived from *deka* = 'hundred' by adding a *d*.

The Gothic form *tehund*, the Greek root *dekad-* and the Lithuanian number word *dešimt* can all be derived from a hypothetical Indo-European substantive

$$*dekṃt = `dekad', \; set \; of \; ten$$

From the last syllable *kṃt* number words like Greek *triakonta* = 'thirty' can be derived. *Triakonta* just means 'three sets of ten'.

To return to Gothic, the usual Gothic word for 'hundred' was not *taihuntehund*, but *hund*. Thus, in the Wulfila bible these number words are used:

200 *tawhunda*
300 *thriahunda*

Obviously, the three words *taihun* = 'ten', *tehund* = 'decad' and *hund* = 'hundred' were closely related. They all contain the essential syllable *hun* and their meanings are directly related to *taihun* = 'ten'. Similarly, the Original Indo-European forms

$$*dekṃ, \; *swkṃt \text{ and } *kṃtom$$

are closely related in form and meaning, and all derived from *dekṃ* = 'ten'. However, it is not known for certain how the ending *tom* in *kṃtom* is to be explained.

The English word *hundred* is derived from the Gothic word *hund*. This was combined with another Gothic word *rathjo* meaning 'reckoning' or 'number' (compare Latin *ratio*). In the Wulfila translation of the story of the loaves and fishes (John 6:10) the words 'five thousand' appear thus:

wairos rathjon swase fimf thusundjos
'men in number almost five thousand'

From *hund* and *rathjo* the word *hund-rathjo* was composed which may be translated as 'hundred in number'. In Old Saxon this word became *hunderod* and in English *hundred*. Thus the history of the word *hundred* can be followed from the Original Indo-European language up to the present.

Figure 36 shows changes in the Original Indo-European words *dekṃ* (= 10) and *kṃtom* (= 100) in the Kentum and Satem languages.

In the quotation from the Gothic bible mentioned above the word *thusundi* (= 1000) was used. In Old Saxon this word became *thusundig*, in Anglo-Saxon *thusund* and in English *thousand*. The Old Norse word was *thushund* or *thushundrad*. Literally this means a 'strong hundred'. The prefix *thus* is derived from an Original Indo-European root *du* which meant something like 'swelling' or 'strong'. From the same root the English word *thumb* (literally 'strong finger') is derived. So

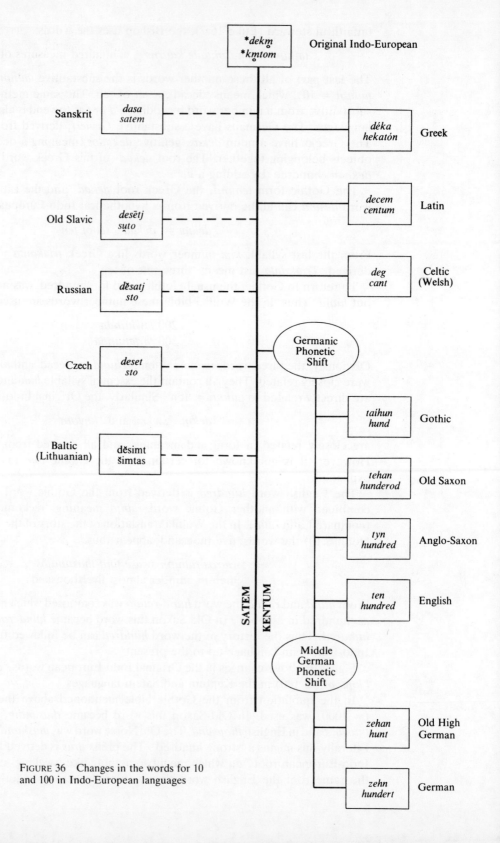

FIGURE 36   Changes in the words for 10
and 100 in Indo-European languages

thousand was originally a 'larger', 'stronger' hundred. There are exact parallels to this in the language of the gypsies; their number words for *ten* and *hundred* are *deš* and *šel*. In Wales, the gypsies say *baro deš* ('great ten') for *hundred* and in Russia they say *baro šel* ('great hundred') for *thousand*.

The Gothic word *thusundi* is not related to the Latin word *mille*. The Original Indo-European number word, from which Latin *mille* and Sanskrit *sahasram* are supposed to be derived, is

$$*sm\text{-}gheslom, \text{ feminine } *smi\text{-}ghsli$$

The middle part of the word, *gheslo* or *gheslio*, became in the dialect of Lesbos *chellio*, in classic Greek *chillioi*. The feminine form *smi-ghsli* was shortened in the ancient Italian languages to *smilli* or *milli* and finally led to Latin *mille*. If these derivations are correct it follows that the Original Indo-European language had a word for 'thousand' already before the Satem languages were separated from the Kentum languages, that is at a very early stage.

Number words like 'thirty', 'three hundred', 'three thousand' are formed in essentially the same way in all Indo-European languages. For 'thirty' there are:

Greek *triakonta*	Sanskrit *trim-sat*
Latin *triginta*	Lithuanian *trys-deszimt*
Tokharian *tary-ak*	Old Church Slavonic *tri-deseti*

For 'three hundred' there are:

Greek *triakósioi*	Sanskrit *tri-śatam*
Latin *trecenti*	Lithuanian *trys-šimtai*
Tokharian *tri-känt*	Old Church Slavonic *tri sta*

For 'three thousand' there are:

Greek *tris-chílioi*	Sanskrit *tri-sahasram*
Latin *tri-milia*	Lithuanian *trys-tukstantis*
Tokharian *tre-wälts*	Old Church Slavonic *tri-trysēšta*

A diagrammatic outline of the derivation of the English word 'thousand' from Original Indo-European may be drawn thus:

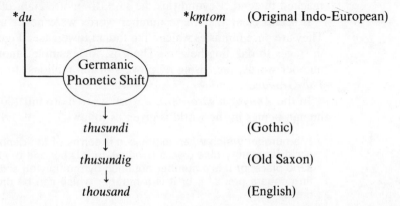

From the evidence of composite number words it is reasonable to conclude that a consistent decimal counting *system* already existed in the third millennium BC in the region where the Original Indo-European language (or languages) was spoken, that is in a region between the Baltic and the Caspian seas.

The non-composite Original Indo-European number words cannot be reconstructed beyond *sṃ-gheslom* (= 1000). In ancient Indo-European languages such as Latin higher powers of ten were expressed by composite number words such as *decies centena milia*, 'ten times a hundred thousands'.

When Homer wanted to express ten thousand, he wrote *deka chillioi*. However, in later times the Greeks had a special word for 10 000, namely *myrioi*. This was derived from an older adjective *myrios*, which meant 'countless'. In the *Iliad* (XXI, 319) Homer says *cherados myrion*, 'pebbles without number'. Perhaps the adjective *myrios* was ultimately derived from *myrmex*, 'ant', because ants are typically countless!

In Sanskrit special names for still higher powers of ten were coined. In the book *Lalitavistara*, a well-known Buddhist work of the first century BC, the Prince Gautama Bodhisattva is reported to have asked for the hand of the daughter of the Prince Dandapani. According to the story Bodhisattva is required to compete with five other suitors in writing, wrestling, archery, running, swimming and arithmetic. He wins with flying colours. In the final test the great mathematician Arjuna asks him:

ARJUNA O young man, do you know the counting which goes beyond the *koti* on the *centesimal* scale?
BODHISATTVA I know.
ARJUNA How does the counting proceed beyond the koti on the centesimal scale?
BODHISATTVA Hundred kotis are called *ayuta*, hundred ayutas *niyuta*, hundred niyutas *kankara*, hundred kankaras *vivara*, hundred vivaras *kṣobhya*, hundred kṣobhyas *vivâha*, hundred vivâhas *utsanga*, hundred utsangas *bahula*, hundred bahulas *nâgabala*, hundred nâgabalas *tiṭilambha*, hundred tiṭilambhas *vyavasthâna-prajñapti*, hundred vyavasthâna-prajñaptis *hetuhila*, hundred hetuhilas *karahu*, hundred karahus *hetvindriya*, hundred hetvindriyas *samâpta-lambha*, hundred samâpta-lambhas *gananâgati*, hundred gananâgatis *niravadya*, hundred niravadyas *mudrâ-bala*, hundred mudrâ-balas *sarva-bala*, hundred sarva-balas *visaṃjñâ-gati*, hundred visaṃjñâ-gatis *sarvajñâ*, hundred sarvajñâs *vibhutangamâ*, hundred vibhutangamâs *tallakṣana*.[10]

The largest number Bodhisattva mentions would be 10^{53}. But he has not yet reached the end. Beyond this, he says, there are eight other series.

It seems clear that these number words were never used for actual counting. They are pure fantasies which, like Indian towers (see Figure 37), were constructed in stages to dazzling heights. There is, for example, another interesting series of number words, increasing by powers of ten millions, to be found in Kâccâyana's *Pali Grammar*.

In the *Anuyogadvâra-sûtra*, a work written around 100 BC, the total number of human beings in the world is given as follows:

...a number which when expressed in terms of the denominations, *koṭi-koṭi*, etc., occupies twenty-nine places (*sthâna*), or it is beyond the 24th place and within the 32nd place, or it is a number obtained by multiplying sixth square (of two) by (its) fifth square (i.e. 2^{96}), or it is a number which can be divided (by two) ninety-six times.[11]

Another very large number is that representing the period known as the *Sîrṣaprahelikâ*. The commentator Hema Candra (eleventh century) states that this number

FIGURE 37   An Indian tower, Qutab
Minar, Delhi (Government of India
Tourist Office, London)

would occupy 194 places. It is also stated to be $8\,400\,000$[28].

Although these very large numbers were 'pure fantasies', there is nevertheless an important development associated with these 'towers' of number words in that, with the development of the concept of place value, the words for the numbers were used to designate the places which the symbol for 1 would occupy in order to represent them in the decimal place value system of numerals. Âryabhaṭa I (fifth century) writes:

> *Eka* (unit), *daśa* (ten), *śata* (hundred), *sahasra* (thousand), *ayuta* (ten thousand), *niyuta* (hundred thousand), *prayuta* (million), *koṭi* (ten million), *arbuda* (hundred million), and *vṛnda* (thousand million), are respectively from *place to place* each ten times the preceding.[12]

The word *million* first made its appearance in Italy, probably in the fourteenth century. Just as *padrone* in Italian is a 'great father', a patron or protector, and just as *salone* is a 'great hall', so *millione* originally meant a 'large thousand'.

It was the merchants of Italy who first used the word *millione* as a specific number word. It appeared in printed texts on arithmetic by the end of the fifteenth century, and subsequently spread via Germany and Holland to other parts of Europe and, eventually, all over the world. It is found in the *Summa de arithmetica, geometrica, proportioni et proportionalita* of the friar Luca Pacioli (1445–1514), published in 1494:

> *mille migliara che fa secondo el volgo el millione*

('A thousand thousands, which is called a million in popular speech').

In the *Triparty en la science des nombres* of Nicolas Chuquet (1445?–1500?), a manuscript written in 1484, there is evidence of Italian influence:

745324·804300·700023·654321 ... le premier point peult signifier million, le second point byllion, ... tryllion, quadrillion, quyllion, sixlion ... nonyllion et ainsi des aultres se plus voultre on vouloit procéder.

('... the first dot indicates a million, the second dot billion, ... trillion, quadrillion, quinquillion, sexillion ... nonillion and so on, as far as one may want to go.')

The words *byllion, tryllion*, etc. seem to have been invented by Chuquet as there is no earlier record of them. In general this sequence of words proceeds upwards in the steps 10^6, 10^{12}, 10^{18}, etc. but in America the word *billion* is used to mean 'thousand millions' (i.e. 10^9). In Europe, the French word *milliard* is used for 'thousand millions'. This word first appeared in the sixteenth century and originally meant 'million millions'. It was in common use in France by the nineteenth century with its present meaning. It did not become well known in Germany, however, until after the peace treaty, signed at Frankfurt-am-Main on 10 May 1871 following the Franco-Prussian War (1870–1), by which France had to cede Alsace-Lorraine and pay reparations of 'five milliard francs'.

From the examples *million* and *billion*, it can be seen how number words are often invented in just one place and then spread from the place of invention over large parts of the earth. Thus, just as the Indo-European number words up to 1000 spread out from one place, somewhere between the Baltic and the Caspian Sea where they were first used, over five continents, so the word *million* spread out from Italy, where it was invented in the fourteenth century, over the whole area of Western civilisation.

Remnants of non-decimal number words

In several languages, there are number words that do not fit into the decimal system. In Chapter 2 mention was made of expressions like *quatre-vingt* and *three score*, which seem to be remainders from an ancient 20-system. There are such expressions in Celtic, Germanic and Romance languages, for example:

Breton: 20   *ugent*
   $40 = 2 \times 20$ *daou ugent*
   $60 = 3 \times 20$ *tri ugent*
   $70 = 10 + 3 \times 20$ *dek ha tri ugent*
   $80 = 4 \times 20$ *thevar ugent*
   $90 = 10 + 4 \times 20$ *dek ha thevar ugent*

Danish: 20   *tyve*
   $50 = 2\frac{1}{2} \times 20$ *halftres*, derived from:
   $60 = 3 \times 20$ *tres* or *tre-sinds-tyve*
   $70 = 3\frac{1}{2} \times 20$ *halvfjers*, derived from:
   $80 = 4 \times 20$ *firs* or *fir-sinds-tyve*
   $90 = 4\frac{1}{2} \times 20$ *halvfems*

Swedish: *fyra sneser* = four *snes* = 80

Portuguese money count:

dois vintens $2 \times 20 = 40$
seis vintens $6 \times 20 = 120$

Sicilian dialects:

du vintini $2 \times 20 = 40$
du vintini e deci $2 \times 20 + 10 = 50$
cincu vintini $5 \times 20 = 100$

Albanian, a Satem language:

20   *nji-zet*
$40 = 2 \times 20$ *dy-zet*

Another non-decimal unit of counting, which is still very popular, is the *dozen*. In Old Norse the *tylft* ('set of twelve') is the most widespread small quantity: 24 was expressed as *tvitylf*, 36 as *thrennartylft*.

Related to the *tylft* is a still larger unit of counting: the *great hundred* (or as it was sometimes called in English *long hundred*) of 10 *tylfts*, or 120 units. It was very popular in Scandinavia and in Iceland. In Iceland, before Christianity was introduced, the word *hundrath* always meant 120 in money calculations as well as in counting military units. After the introduction of Christianity, a distinction was made between the 'hundred in 10-count', which was used in the Bible and in learned writings, and the 'hundred in 12-count'.

In a manuscript dated *c.* 1250 there is the phrase

$$III^c \; daga \; tolfroed$$

which means 'three hundred days by the 12-count', and refers to the 360 days thought to constitute the year. The writer goes on to comment that in Latin (referred to as 'the book language') all hundreds are reckoned according to 10-count.

It seems that the 'great hundred' was considered not as a dozen tens but as ten dozens for in Lübeck in Northern Germany people counted *Hundert Bretter* as *10 Zwölfter* ('10 twelves') or, as they said in Mecklenburg, *10 Tult*. (The word *Tult* is related to Old Norse *tylft*, which means 'dozen'.)

However could the *great hundred* be related to vigesimal or sexagesimal number words, i.e. is there any linguistic evidence for $120 = 6 \times 20$ or 2×60 in the Indo-European languages? This question can be answered fairly confidently in the negative. While there is evidence that the number words for 60 have played a special role in cultural history[13], in no case is there a number word corresponding to *great hundred* composed of words for 2 and 60 nor is the number 120 of special significance in any sexagesimal or vigesimal system.

4 Written Numbers

The discussion so far has been concerned almost entirely with the history of *spoken* numbers. This chapter is concerned solely with the history of *numerals, i.e. written numbers*. It describes numeral systems which have been used in different parts of the world and traces the familiar decimal numbers back to their origin in the Indian subcontinent.

Egyptian numerals

The oldest known numeral systems are those of the ancient Egyptians and those of the Sumerians in Southern Babylonia. The Egyptian system will be discussed first, as it is easier to describe and in some respects more primitive than the Sumerian system.

The earliest Egyptian system, *hieroglyphic* numerals, is found on inscriptions dating back as early as 3000 BC. It is based on the *repetition* of symbols for one, ten, hundred, thousand, ten thousand, hundred thousand, and million (see Figure 38).

Hieroglyphs were written from right to left so the number 246, for example, appears as: III ∩∩ ϑϑ

1	I	‖‖	6
2	I I	‖‖	7
3	III	‖‖	8
4	IIII	‖‖	9
5	‖‖		

10	∩	
100	ϑ	
1 000	𐦜	Lotus
10 000	𓂭	
100 000	𓆐	Tadpole or bird
1 000 000	𓁨	Not in later use

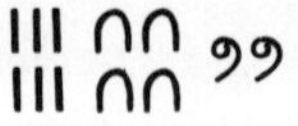

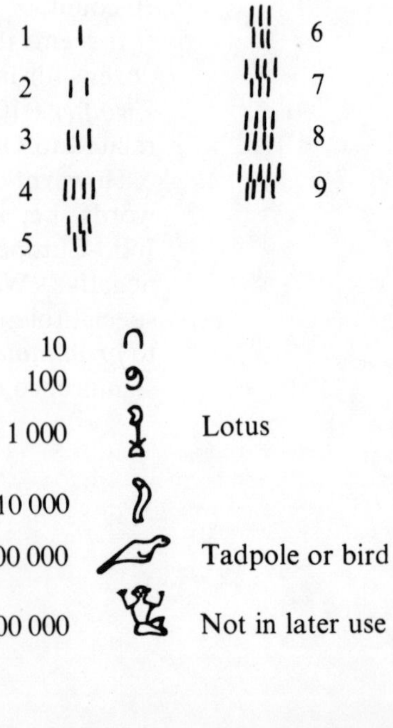

FIGURE 38   Egyptian symbols for 1 to 9 and large numbers

Because there were different symbols for one, ten, hundred, etc., no great confusion would arise if numerals of lower value were written before instead of after those of higher value. Thus ⟨99 ∩∩ |||⟩ represents 246 quite unambiguously despite the reversal of the order of writing down the numerals. Also, there is no need for a zero: 2018, written as ⟨|||| ∩ 99⟩ cannot be confused with, for example, 218 written as ⟨|||| ∩ 99⟩ . This is in sharp contrast with our system of numerals where 246 cannot be read as 642, and 218 has to be distinguished from 2018 by writing a zero in the hundreds place.

Only a limited amount of mathematical information can be gleaned from the hieroglyphic inscriptions. Present knowledge of how calculations were performed with these Egyptian numerals derives from a number of surviving papyri, and in particular from the *Rhind* or *Ahmes Papyrus* (see Figure 39), a roll about a foot high and some eighteen feet long, originally written sometime after 1800 BC[1]. The roll is written in the more cursive hieratic script; this can, however, be readily transcribed into hieroglyphs.

To multiply 12 by 12 the Egyptians would first double the number 12 and then double the result. This would give

$$4 \times 12 = 48$$

Doubling this again would give

$$8 \times 12 = 96$$

and by adding 48 and 96, the final result of 144 would be obtained. This calculation is shown below in hieroglyphs and also in modern transcription.

	1	12
	2	24
	/4	48
	/8	96 sum 144

To proceed more rapidly the Egyptians frequently multiplied by 10; also they sometimes halved the results obtained. Thus in the Kahun papyrus multiplication of 16 by 16 is performed as follows:

	/1	16
	/10	160
	/5	80 sum 256

FIGURE 39   Extract from the Rhind Papyrus (British Museum)

Multiplying 110 by 15 by doubling, and by using multiplication by 10 is shown below. Multiplication by doubling:

∩૧	⏐ ⌐	/1	110
∩∩૧૧	‖ ⌐	/2	220
∩∩∩∩૧૧૧૧	‖‖ ⌐	/4	440
∩∩∩ ૧૧૧ ۹ ⌷ ∩∩∩∩૧૧૧૧	‖‖‖ ⌐	/8	880 sum 1650

Multiplication by ten:

∩૧	⏐	1	110
૧૪	∩⌐	/10	1110
∩∩∩ ૧૧૧ ۹ ⌷ ∩∩∩૧૧૧	‖‖ ⌐	/5	550 sum 1650

Because of the common practice by historians of transcribing the hieratic texts, such as the Rhind Papyrus, into hieroglyphs for purposes of discussion, a false impression is given of the relative importance of hieroglyphic and hieratic numerals. Professor Boyer writes:

Such an impression unfortunately is one which is allowed by histories of mathematics to remain. The alternative hieratic (and demotic) forms customarily are not mentioned or are referred to only casually, as though they were seldom used or illustrated no new principle, and hence warranted only antiquarian attention. This constitutes a serious misrepresentation of Egyptian notation and of the fundamental principles of numeration. The hieroglyphic forms appear to have been reserved largely for formal inscriptions, whereas even the oldest papyri, such as the Rhind and Moscow, made use of the hieratic. Furthermore, the former are typical of a more primitive stage in numeration, whereas the latter represent an advance of the greatest importance. The Egyptians early recognized that the ancient hieroglyphic script could, through the use of conventionalized symbols, be considerably simplified and abbreviated, and so this form of writing began to give way even in the Old Kingdom to the more cursive hieratic and later still to the demotic. These paleographic changes in Egypt resulted incidentally in a totally new basis for numeration. Scribes abbreviated numerical representations by introducing new marks which were far more concise and yet adequately distinctive. This took the form of replacing a collection of symbols by a single mark or cipher, as is indicated in the accompanying table [Figure 40]. In general, groups of iterated number signs

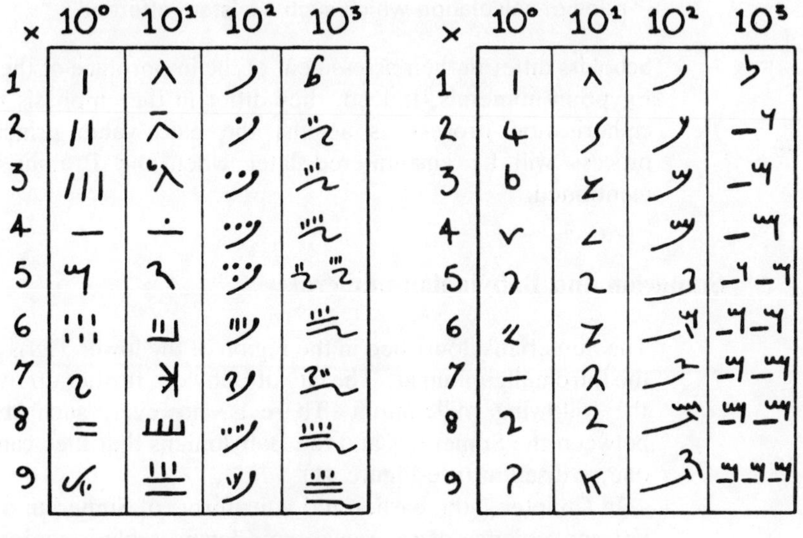

FIGURE 40 Egyptian ciphered numerals, hieratic and demotic forms

were avoided by substituting for them new characteristic symbols. Such a scheme, if carried out completely, would necessitate a distinctive mark or 'cipher' for each of the first nine natural numbers and for each of the first nine integral multiples of integral powers of ten. Such a system of numeration might well be called a decimal cash-register cipherization. In it any number less than one thousand would be represented by not more than three symbols—one for units, one for tens, and one for hundreds—just as any sum less than $10.00 is indicated on most types of cash-registers by not more than three keys or flags. Such a change, however, calls for a great multiplicity of new number symbols. Thus fifty-four symbols would be needed for the complete decimal cash-register cipherization of the integers less than one million. It would be interesting to speculate on the part which the flexibility afforded by Egyptian writing materials may here have played in making available new symbols with which to establish the principle of cipherization, and to compare this with the extent to which in Mesopotamia a characteristic inflexibility, with a corresponding need for an economy of symbols, may have operated to lead to the other great principle of numeration, local value.[2]

The introduction by the Egyptians of the idea of cipherization constitutes a decisive step in the development of numeration, and their contribution in this connection is in every way comparable in significance to that of the Babylonians in adopting the positional principle. The larger number of symbols called for in a ciphered system imposed a greater tax on the memory, but Egyptian scribes evidently regarded this as justified by the greater ease in reading and writing, as well as by an increased facility with respect to tabular arrangements in calculation. Nevertheless, the Egyptians did not exploit their invention to the full. In the first place, the hieratic script was not completely ciphered: a few of the digits, such as two, three, and six (and sometimes also four), continued to be represented by repeated strokes or marks; for others, especially the hundreds and thousands, multiplicative and additive principles were adopted. Then too, some of the ciphers, such as for ninety and for five hundred, were unnecessarily intricate, while others, especially in the demotic forms, were not sufficiently distinctive. Furthermore, the forms varied considerably from scribe to scribe, so the system was not comparable in universality to our own as used in Western civilization. Then, too, the system was not systematically extended to include indefinitely large numbers beyond thousands. Moreover, the Egyptians appear not to have made effective use of the rules of calculation which such a system afforded.[3]

Scholars differ in their assessment of the importance of the cipherisation process in Egyptian numerals. Indeed, they differ in the emphasis which they place on the cipherisation process as against the place-value principle. The cipherisation process will be encountered later when the Brâhmî numerals in India are mentioned.

Sumerian and Babylonian numerals

The Sumerians flourished in the region of the lower Tigris and Euphrates valleys in the third millennium BC. The city of Babylon, further up river, became important in the following millennium. There is, however, such strong cultural continuity between the Sumerians and the Babylonians that they can be treated as phases of one civilisation (see Figure 41).

In Chapter 2 the earlier and later forms of Sumerian numerals were described. For convenience of reference the relevant table is represented in Figure 42. The symbols were made by using soft clay and a stylus, as shown in Figure 43. The original forms were made using a round stylus and the later (cuneiform) forms using a triangular prismatic stylus.

In the original form (in use around 3000 BC) the symbols for 1 and 60 could be distinguished, the latter being larger. With the later (cuneiform) numerals, 1 and 60 were both represented by a simple wedge. Sometimes the wedge for 60 was made a little larger than that for 1, but in most cases no difference can be seen. So the question arises: how could the Sumerians and the Babylonians distinguish between 1 and 60 when reading a cuneiform text? The answer to this is that sometimes the distinction could be made from the position occupied by the wedge, and sometimes from the context in which the number represented occurred.

For example, if one reads in a text the whole number ❚ ⟨⟨⟨ it is clear that the first wedge represents 60 and not 1, since longer units always precede smaller ones (and the Sumerians, unlike the Egyptians, wrote from left to right). Again with the subtraction ❚ minus ⟨⟨ gives ⟨⟨⟨⟨ , it is clear that the simple wedge means 60 and not 1. Thus, in mathematical contexts, there is generally little difficulty in making the correct distinction. In other cases the distinction has to be

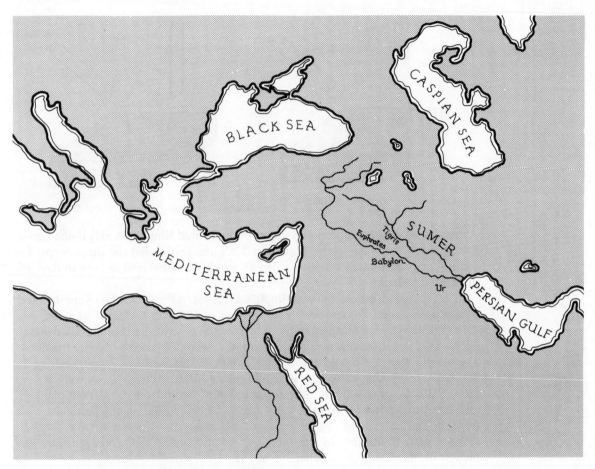

FIGURE 41   Sumer and Babylon

Value		Original form	Cuneiform	Sumerian number word
1		D	![Y]	*aš*
	10	o	![<]	*u*
60		D	![Y]	*geš*
	60 · 10	![D]	![K]	*geš-u*
		O	![star]	*šar*
60²				
	60² · 10	![O]	![star]	*šar-u*
60³				*šar-gal*

FIGURE 42   Early Sumerian numerals

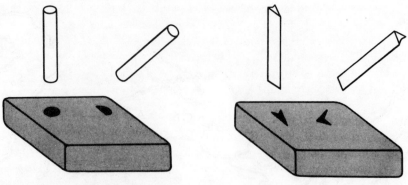

FIGURE 43   Styli and clay tablets

inferred from the general context, just as in the United Kingdom today if one sees a pair of slippers and an orange both priced '8', one knows that the slippers cost £8 and not 8p, and that an orange costing £8 would be unbelievable even in days of substantial inflation.

In later times, the notation was simplified further. All powers of 60 were denoted by a simple wedge **Y** , and ten times a power of 60 by an L-shaped wedge **<** .

In Old Babylonian mathematical texts from the time of the Hammurabi dynasty (1830–1531 BC), the same symbols, **Y** and **<** , were also used to denote fractions. The simple wedge **Y** could mean $1/60$, $1/60^2$, ... as well as ..., 60^2, 60, or 1. Similarly, the L-shaped wedge **<** could mean $10/60$, $10/60^2$, ... as well as ..., 10.60^2, 10.60, or 10. Thus, for example **Y<<** could mean 80 multiplied or divided by any power of 60. The value had to be determined in any particular case from the context.

In modern transcriptions the following notation (due to O. Neugebauer) has proved useful. Take the number **Y Y < YY** .

If the value of the wedges is not known, this is transcribed as 2,15, that is, $2 \times 60 + 15$ (= 135) multiplied or divided by any power of 60. If it is known that the first two wedges each signify 1, then this is 2;15 – the semicolon being used to denote the separation between the integral and fractional parts of the number. Thus 2;15 means $2 + 15 \times 1/60$ (= $2\frac{1}{4}$). If the first two wedges each have the value 60^2, then one writes 2, 15, 0, which means $2 \times 60^2 + 15 \times 60$ (= 8100).

The Babylonian system of writing numbers is a *place-value* system – the value of a wedge depends on its position. To show this even more clearly, consider the number **Y <<Y**. The *places* of the various wedges can be shown as follows (assuming that their values have already been determined):

60	10	1
Y	**<<**	**Y**

The first wedge has a value sixty times that of the last one.

The systematic use of a place-value notation has great advantages in computations, and the Babylonian *base* of 60 has the special advantage that it has so many factors. Many divisions can be directly reduced to multiplications. Consider, for example, the division of 4 by 15. This would be a difficult operation for the Egyptians. For the Babylonians it was very easy: 1/15 is equal to 4/60, so 1/15 was written the same as 4, and 4 times 4 is 16. In Neugebauer's notation this is 4 times 0;4 equal to 0;16. Division by numbers having only the factors 2, 3 or 5 could always be performed in this way. In other cases the Babylonians resorted to approximations.[4]

A serious drawback of ancient Babylonian notation was that it was not possible to distinguish between, for example, 1,0,30 and 1,30 because there was no symbol for zero. To overcome this difficulty later Babylonian scribes introduced a special double-L-shaped wedge to indicate where there was an empty place between two digits. Thus 𒁹𒑱𒀭 = 1,0,4 = 3604. However, this zero symbol does not seem to have been used at the end of a number; so in fact, 𒁹𒑱𒀭 could mean 3604 multiplied by any power of 60.

The sexagesimal place-value system was used not only by the scribes of the Old Babylonian period, but also by later Babylonian astronomers. They divided the zodiacal circle[5] into twelve signs of the zodiac of 30 degrees each. The degree was divided into 60 minutes, and the minute into 60 seconds. This system was later adopted by the Greek astronomers[6], and is still in use today. The Alexandrian astronomer Ptolemy (second century AD), and possibly some of his predecessors, used the symbol 'o' for zero not only between digits but also at the end of a number. This gave the 'finishing touch' to the sexagesimal system – the value of every written number was now completely determined.

This discussion has considered Sumerian and Babylonian numerals only after the clear establishment of the sexagesimal system, a system which developed out of earlier systems of mensuration. From recent evidence obtained from earlier tablets, usually referred to as 'proto-Sumerian', it appears that there were several different systems of conversion factors in use in earliest times and that these differed according to the particular kind of measurement being undertaken: for example, the factors for horizontal linear measurement were not the same as those for height, though all would divide exactly into 60. A typical example (for linear measurement) is:

6 *shi*	= 1 *shu-si*
30 *shu-si*	= 1 *kush*
12 *kush*	= 1 *nindan*
60 *nindan*	= 1 *ush*
30 *ush*	= 1 *beru*

In some cases the conversion factor 10 was also used. Note that all these factors divide exactly into 60, a far more convenient base than the base 10 which is not sufficiently large and does not have the appropriate factors for the construction of a family of nicely interrelated measurement systems. Compare, for example, the succeeding factors involved in measuring from the inch to the mile: 12, 3, $5\frac{1}{2}$, 4, 10 and 8.

The usefulness and potential of the Babylonian and Egyptian systems for writing numbers can be briefly compared. The Egyptian system is more primitive than the Babylonian system in that it is not a place-value system. It is more difficult to perform calculations in the Egyptian system. The Babylonian system lacked a symbol for zero, which was not developed until fairly late, so that for just writing down numbers the Egyptian system had the advantage of being unambiguous.

The Babylonian place-value system required only two symbols, since the place-value principle obviates the necessity of inventing new symbols for large numbers and fractions. The ability of the place-value system to represent fractions in the same way as whole numbers is particularly significant – there is no theoretical limit to the accuracy with which calculations may be carried out.

Chinese numerals

There are five distinctive kinds of Chinese numerals (see Figure 44): the *standard* (or *basic*) numerals, the *official* numerals, the *commercial* numerals, the *stick* (or *rod*) numerals and the *oracle-bone* numerals. All of these have their origins in ancient times and all are decimal.

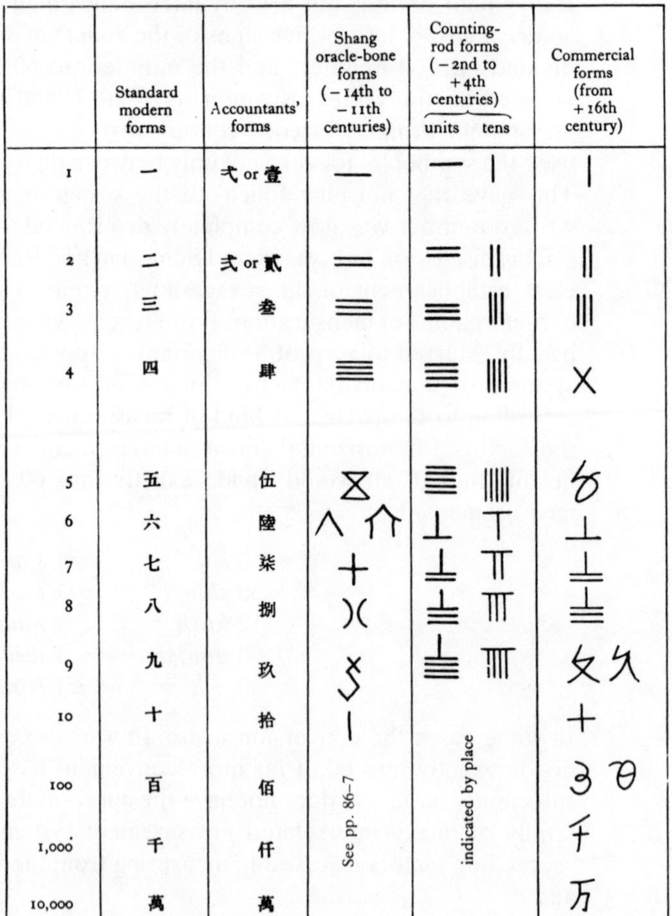

	Standard modern forms	Accountants' forms	Shang oracle-bone forms (−14th to −11th centuries)	Counting-rod forms (−2nd to +4th centuries) units / tens		Commercial forms (from +16th century)
1	一	弍 or 壹	—	—	丨	丨
2	二	弎 or 貳	=	=	‖	‖
3	三	叄	≡	≡	⫴	⫴
4	四	肆	≣	≣	⫼	✕
5	五	伍	☒	≣	⫿	⅄
6	六	陸	∧	⟰	⊥	⊥
7	七	柒	+	丄	丅	丄
8	八	捌	)(	⫤	𝍤	⫤
9	九	玖	⌇	⫥	𝍦	夂久
10	十	拾	丨			十
100	百	佰				�$ 𝟋
1,000	千	仟	See pp. 86–7	indicated by place		千
10,000	萬	萬				万
0	零	零				○

FIGURE 44   Chinese numerals (by courtesy of Cambridge University Press)

四   4
萬   TTh
一   1
千   Th
九   9
百   H
五   5
十   T
七   7

FIGURE 45 Chinese numeral for 41957 in standard numerals (by courtesy of MIT Press and Vandenhoek & Ruprecht, Göttingen)

The standard numerals are called *hsiao-hsieh*, meaning the 'small (or common) writing', and they are often referred to as the 'modern forms' though they are ancient and mediaeval as well. With the exception of the zero, they have been in use in the form shown since the third century BC. The system is decimal but it is not a place-value system. It is, however, superior to the Egyptian and Roman systems. When the Romans wanted to write 400, for example, they had to repeat the symbol C for 100 four times, thus: CCCC, whereas the Chinese write the word *szu* (= 4) and below it the word *pai* (= 100), thus:

四 *szu* = 4
百 *pai* = 100

The writing is in vertical columns from top to bottom. To write out a number the Chinese only have to write in one column the single number words in the order in which they are spoken. For example, the number 41957 is written as shown in Figure 45. In words this is *szu-wan i-ch'ien chiu-pai wu-shih ch'i*.

The *official* numerals are called *ta-hsieh*, meaning the 'great writing'. Their highly ornamented forms are used wherever a number needs to be protected from falsification: on banknotes, contracts, bank drafts, etc. (see Figure 46). They came into use from the first century BC onwards, but are naturally found only very occasionally in mathematical texts.

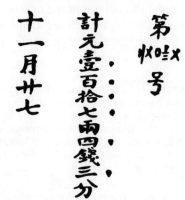

FIGURE 46 Chinese bank draft with three sets of numerals: basic, official and commercial. First column (right), the check number 24 084 in commercial numerals; second column, the amount 117.43 taels in official numerals; third column, in basic numerals 'the 27th day in the 11th month' with the 20 in the old form (by courtesy of MIT Press and Vandenhoek & Ruprecht, Göttingen)

The *commercial* numerals are used by merchants and tradespeople to write numbers rapidly on less critical documents such as price tags (see Figure 47). They are what a present-day visitor to China would probably find on a restaurant bill. They are called *ma tzu* or *an ma tzu*, meaning 'confidential weight numerals', and date back to the sixteenth century, though they incorporate some numerals dating from ancient times.

The main concern here is with the *stick* and *oracle-bone* numerals. The *stick* numerals are formed out of horizontal and vertical strokes. The vertical strokes represent units or hundreds or ten thousands, etc., the horizontal strokes tens or

FIGURE 47 Chinese commercial price tag: 72 (liang) 5 (chi'en) 3 (fen) 2 (li) 725 (chi'en) 3 (fen) 2 (li) (by courtesy of MIT Press and Vandenhoek & Ruprecht, Göttingen)

thousands, etc. (see Figure 48). These numerals go back to at least the second century BC and originate from arrangements of calculating sticks laid out on a counting board (see Chapter 6). Their present form was stabilised in the thirteenth century. A circular zero then indicated missing places. For example, $| \equiv \Pi\ O\ O\ O\ O = 1\,470\,000$.

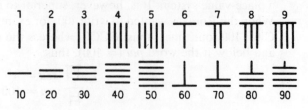

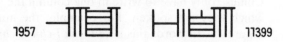

FIGURE 48   Examples of stick numerals, made up of vertical and horizontal strokes; within a given number the orientations alternate (by courtesy of MIT Press and Vandenhoek & Ruprecht, Göttingen)

Numerals identical to the stick numerals can be found in other types of Chinese systems for writing numbers, though not in the ornate official forms. For example, the commercial forms for 1, 2, 3, 6, 7 and 8 are identical to the corresponding stick forms. The stick system is a place-value system. Of the zero symbol Joseph Needham writes:

> The circular symbol for zero is first found in print in the *Su Shu Chiu Chang* of Chhin Chiu-Shao (+ 1247), but many have believed that it was in use already during the preceding century at least. The usual view is that it derived directly from India, where it first appears on the Bhojadeva inscriptions at Gwalior dated +870c. But there is no positive evidence for this transmission, and the form could perhaps have been borrowed from the philosophical diagrams of which the +12th-century Neo-Confucians were so fond.[7]

The *oracle-bone* numerals date from the fourteenth to the eleventh centuries BC, and very similar forms have been found on coins dating from the tenth to the third centuries BC (see Figures 49 and 50). The horizontal strokes in the signs for 100 and 1000 have the meaning 'one', so the symbol for 100 means 'one hundred' and the symbol for 1000 means 'one thousand'.

To write the number 300 or 3000 the stroke had to be replaced by three strokes. These symbols were combined for the purposes of recording large numbers; for example:

56 图图图 *wu-shih yu lui* 'five tens plus six'
656 图图图 *liu-pai wu-shih lui* 'six hundreds, five tens, six'

This system can be thought of as a *named place-value system*. As in our place-value system, only ten numerals were needed (in our system the numerals from 0 to 9; in the Chinese system those from 1 to 10), but because the zero was missing, special symbols like 图 or 图 were needed to indicate whether the numeral 6 stood for 6 hundreds or 6 thousands.

FIGURE 49
Oracle-bone numerals for 100, 1000, 300, 3000 (by courtesy of Cambridge University Press)

FIGURE 50 Shang oracle bone, about 1300 BC. In part, the inscription reads: 'On the seventh day of the month, . . . a great new star appeared in company with Antares . . .' (by courtesy of Cambridge University Press)

A symbol for zero in the form of a dot is mentioned in the *Khai-Yuan Chan Ching*, a great compendium of astronomy and astrology edited between AD 718 and 729. Chapter 104 of this work contains a section on Indian methods of calculation. After saying that numerals are all written cursively with only one stroke each, the writer states:

> When one or other of the 9 numerals is to be used to express a multiple of 10, then it is entered in a column in front of the unit digit. Wherever there is an empty space in a column, a dot is always placed.[8]

According to Needham, a dot representing zero is also found on an inscription written in Cambodia in AD 603. On this inscription, the year 605 of the Saka era (= AD 603) is represented by ς·ξ . Soon afterwards (AD 686) a little circle was used as a symbol for zero on the island of Banka in Indonesia. The symbol for zero was probably invented in India together with the decimal place-value system in the sixth century AD. After its invention, it quickly spread to Indo-China, to Indonesia and to China.

Greek numerals and their derivatives

The earliest numerals found in the area of Greek civilisation, for instance in Crete around 1000 BC, seem to be a fairly random collection of number symbols, considerably more primitive than the systems then in use in Egypt or Mesopotamia. The Greek *numeral system*, known as the *Attic* (or *Herodianic*) *numerals*, probably dates from around 500 BC (see Figure 51). The system is basically repetitive, like the Egyptian hieroglyphic system. Thus 23, for example, was written ΔΔIII. The symbols Γ (or Π), Δ, X, M are respectively the first letters of the Greek number words ΠΕΝΤΕ (5), ΔΕΚΑ (10), ΧΙΛΙΟΙ (1000), ΜΥΡΙΟΙ (10000). ʽΕΚΑΤΟΝ (100), however, begins with a short E and not H (êta).

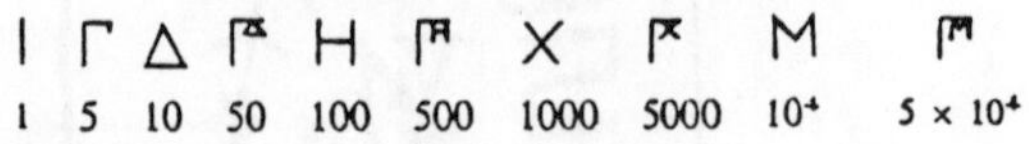

I	Γ	Δ	Γᐞ	H	Γᴴ	X	Γˣ	M	Γᴹ
1	5	10	50	100	500	1000	5000	10^4	5×10^4

FIGURE 51   Early Greek numerals

The system was similar to the well-known Roman system in using letters, but it had one important advantage. The Roman system had distinctive symbols for 50 and 500, but in the Attic system these were written by combining the symbol Γ for 5 with those for 10 and 100, so that 50 becomes Γᐞ – a combination of Γ and Δ; and 500 becomes Γᴴ – a combination of Γ and H . This principle was also extended for 5000 and 50000. The system, though decimal, was rather cumbersome and gradually a briefer alphabetic notation was more widely adopted.

1–9	$\alpha, \beta, \gamma, \delta, \varepsilon, \varsigma, \zeta, \eta, \theta$	(6 = ς = vau)
10–90	$\iota, \kappa, \lambda, \mu, \nu, \xi, o, \pi, \varsigma$	(90 = ς = koppa)
100–900	$\rho, \sigma, \tau, \upsilon, \phi, \chi, \psi, \omega, ℸ$	(900 = ℸ = sampi)
1000–9000	α, β, etc.	

FIGURE 52   Ionian numerals

In this *Ionian* system (see Figure 52) 27 symbols were needed. Since the usual Greek alphabet in later times had only 24 letters, three ancient letters *digamma* or *vau*, *koppa*, and *sampi* were used to denote *six, ninety,* and *nine hundred* respectively. The old form of the letter digamma or vau was Ϝ, but the later form was Ϛ, which was also called *stigma*, because it served as an abbreviation for *st*. Numbers were distinguished from words by writing an accent (prime) at the end or by putting a line over the letters, thus:

$$\alpha\tau\varepsilon' \text{ or } \overline{,\alpha\tau\varepsilon} = 1305$$

Numbers greater than the myriad M = 10000 were written either by writing numeral letters above the symbol M or by placing dots above the letters, thus:

$$\overset{\kappa\varepsilon}{M}\mu\gamma' \text{ or } \ddot{\kappa}\ddot{\varepsilon}\mu\gamma = 250\,043$$

Notice that no symbol for zero is required.

Archimedes (third century BC) invented a notation for higher powers of M by means of which all numbers, no matter how large, could be expressed. In his treatise the *Sand-Reckoner* he boasted that he could write down a number greater than the number of grains of sand needed to fill up the Universe.

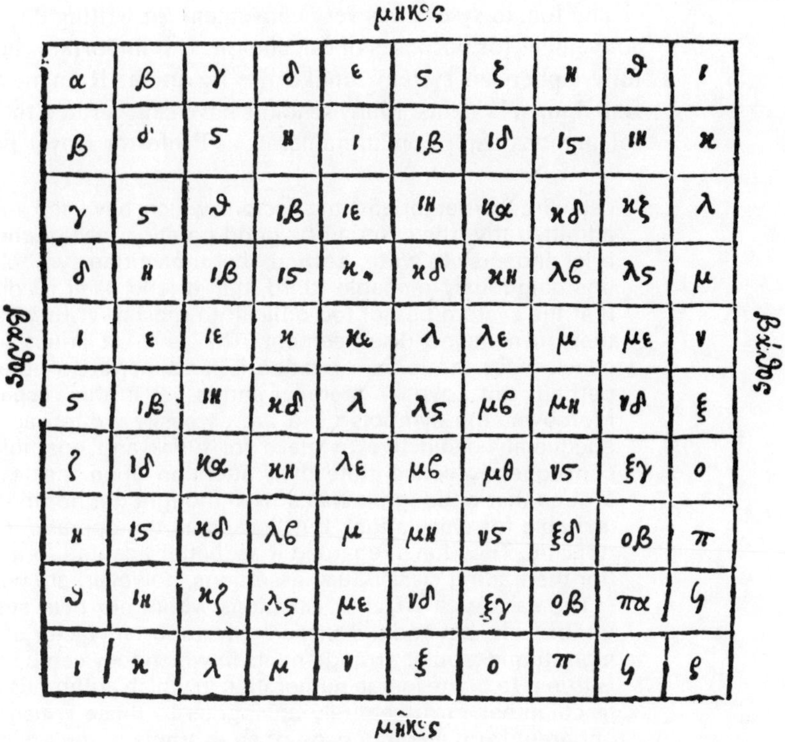

μῆκος

α	β	γ	δ	ε	ς	ζ	η	θ	ι
β	δ	ς	η	ι	ιβ	ιδ	ις	ιη	κ
γ	ς	θ	ιβ	ιε	ιη	κα	κδ	κζ	λ
δ	η	ιβ	ις	κ	κδ	κη	λβ	λς	μ
ε	ι	ιε	κ	κε	λ	λε	μ	με	ν
ς	ιβ	ιη	κδ	λ	λς	μβ	μη	νδ	ξ
ζ	ιδ	κα	κη	λε	μβ	μθ	νς	ξγ	ο
η	ις	κδ	λβ	μ	μη	νς	ξδ	οβ	π
θ	ιη	κζ	λς	με	νδ	ξγ	οβ	πα	ϟ
ι	κ	λ	μ	ν	ξ	ο	π	ϟ	ρ

μῆκος

FIGURE 53 Extract from a greek multiplication table of Nichomachus of Gerasa, 1538 (Turner Collection, University of Keele)

The Greek method of calculation using the Ionian numerals (see Figure 53) was very similar to our own. For example consider the multiplication 265×265, as worked out by Eutocius in his commentary on the works of Archimedes. Modern decimal notation begins by multiplying 5 by 5. The Greeks would similarly begin by multiplying ϵ by ϵ, having first decomposed $\sigma\xi\epsilon = 265$ into $\sigma + \xi + \epsilon$. The single terms would be multiplied and the results added. But how would the Greeks have multiplied σ by ξ (i.e. 200 by 60)? The modern system merely multiplies 2 by 6 and writes the result 12 in the correct places – an almost automatic procedure. But the Greeks had first to go back from $\sigma = 200$ and $\xi = 60$ to the 'root numbers' $\beta = 2$ and $\varsigma = 6$. They were then able to calculate as we do $2 \times 6 = 12$, i.e. β times $\varsigma = \iota\beta$. Before the result could be written down, it had to be multiplied by $100 \times 10 = 1000$ giving the partial result $\overset{\alpha}{M},\beta$. Adding all the partial results obtained in this way gave the final answer (see below). This method of multiplying was known as *Greek multiplication* to distinguish it from the Egyptian method of repeated doubling which was known to the Greeks.

τὰ δὲ $\overline{\sigma\xi\epsilon}$			
ἐπὶ $\overline{\sigma\xi\epsilon}$			
$\overset{\delta}{M}\overset{\alpha}{M},\beta,\bar{\alpha}$			
$\overset{\alpha}{M},\beta,\overline{\gamma\chi\tau}$			
$\overline{,\alpha\tau\kappa\epsilon}$			
ὁμοῦ	$\overset{\zeta}{M}\overline{\sigma\kappa\epsilon}$		

i.e.	265		
times	265		
	40 000	12 000	
		12 000	3 600	300
		1 000	330	25
together	70 225		

The Ionian sysem was very convenient for writing down numbers, but much less convenient for purposes of calculation. It is important, however, to note that it is a fully cipherised system, unlike the Egyptian, Roman, and even the place-value Babylonian systems. Many scholars have exaggerated the difficulties of performing calculations using Ionian numerals as Professor Boyer points out[9]:

Let us consider briefly the factors which have led to such criticism. It will be admitted that the criteria of a good notation include the following: first, that it be brief and easy to write; second, that the written symbolism be quickly, easily, and unambiguously readable; third, that it lend itself readily to computation; fourth, that the system be not too difficult to master. With respect to the first two points there can be no disagreement. The severest critics are quick to admit that in conciseness the Ionian notation is far superior to other ancient systems (except perhaps the Egyptian hieratic) and is not in this respect surpassed by our own. Moreover, the symbols used are as easily written as any ever devised and are adequately distinctive to make for quick and unambiguous recognition. Critics consequently concentrate their attention upon the last two criteria. Many have argued that although it was a 'well thought out homogeneous system', it was not 'suitable for calculation', but 'makes all the operations of arithmetic exceedingly difficult'. They have regarded it as 'better adapted to the theory of numbers' than 'for the trading class'. Such assertions, however, are without any basis in fact...

The ease with which calculations within our own system can be carried out is attributable, beyond the exceedingly important factor of conciseness, largely to the algorithmic tabular arrangements in which they can be represented. However, this is true also of the Ionian numeration in which arithmetic processes were carried out in columnar forms entirely analogous to those which we ourselves use. This is apparent from a comparison of an example of multiplication taken from Eutocius' commentary on the *Measurement of the Circle* of Archimedes with its modern equivalent. In this illustration the Greek example contains more figures than are necessary, for digits are written down which now are normally carried mentally [see Figure 54]. Such mental effort is more called for in our system inasmuch as confusion in place or column is fatal ... Such confusion does not, of course, occur in the Ionian notation, and this fact must be recognized as a very definite advantage over our own method of numeration. However, one may, if he wishes, carry figures mentally in the Greek system as well as in ours.

FIGURE 54   Greek multiplication

It is conventionally maintained that the tabular arrangement in which calculations today are carried out with such ease is the direct result of the principle of position. That such a view is without any foundation is apparent from an illustration. In the accompanying chart [see Figure 55] the product of 4506 by 23 is represented tabularly in terms of four systems of numeration: the iterative Egyptian hiero-glyphic, the iterative positional Babylonian, the ciphered Ionian, and the ciphered positional 'Hindu-Arabic'.

A consideration of these four forms will show that calculation is facilitated not so much by a positional *notation* as by a positional *arrangement*—i.e., by an ordered tabularization corresponding to the manner in which algebraic polynomials are customarily written. Such a tabular arrangement is possible in *any* system of numeration. Nevertheless, it will be clear that two of the above systems in this

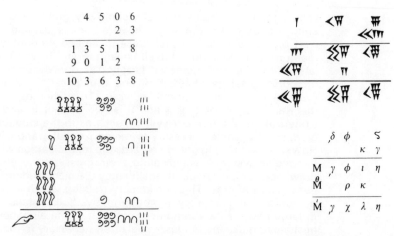

FIGURE 55   Examples of Greek, Egyptian, Babylonian and Hindu-Arabic multiplication

respect afford distinct advantages over the others. The Babylonian and Egyptian schemes are equally clumsy, yet one does, the other does not, employ local value. The Babylonian is, in fact, often more awkward than the non-positional Egyptian. This *greater* clumsiness, however, is not to be ascribed to the principle of position, but rather to the confusion engendered among the Babylonians by the incongruous use of decimal auxiliary stopping-points in a system otherwise logically sexagesimal. British and American civilizations continue to make the old Babylonian mistake of confusing two (or more) scales of notation. Duodecimal aggregates such as the dozen, the shilling, feet of twelve inches, and days of twice twelve hours are hopelessly out of place in a civilization which has dedicated itself in other connections to the ten-scale. The decimal system is not, of course, better than the duodecimal; if anything, it is perhaps slightly less convenient. The radix of a ciphered number scale should in the first place be not too large or too small. Neugebauer has overlooked the fact that the Babylonian sixty is far too large[10]; the binary radix[11], praised by Leibniz, is far too small. Eight, ten, and twelve are convenient in size. However, ten is not integrably divisible by either three or four and for this reason is less desirable than either eight or twelve. However, the choice of one of these to serve as the modulus is of less consequence than that the system of numeration shall be homogeneous, based consistently upon one and only one radix...

The Babylonian, the Mayan, the Herodianic, and the Roman all failed to observe this rule; only the Egyptian and the Ionian were consistent in this respect. And consistency is essential for ease of calculation, whether the system of notation be positional or ciphered.

On the surface, at least, the last two forms in the illustrations above are far superior, yet here again only one of these is positional. It is significant, however, that each of the last two is ciphered. It should be clear from such a comparison that the adaptability of a system of numeration to tabular arrangements in computation is facilitated by a cash-register cipherization. Under no circumstance is it to be attributed solely or even largely to the principle of place value.

The examples above in each case have been set out in the most favorable form, compatible with the principles of the system, which these peoples might reasonably have been capable of developing. That the Alexandrian Greeks, in the few examples of logistic which have come down to us, did arrange their operations in columns more or less as indicated would in itself argue for cipherization as a leading factor in the development of such tabularization.

The one possibly legitimate objection to be raised against the alphabetic notation arises under the fourth point. It is argued that 'the mental effort of remembering such an array of signs was comparatively great', and that it involved 'a most annoying tax on the memory'. However, one is justified in taking such statements *cum grano salis*. In the first place, a limited amount of necessary preliminary memorization does not of itself constitute an objection, for it may offer rewards later in the form of increased facility. We ourselves are faced with the need for memorizing tables of addition and multiplication which are not called for in systems of the more primitive types. To earlier peoples it had been obvious that ⦀ and ⦀ is ⦀⦀ , or that ⦀⦀ and Γ is Γ⦀⦀ , or that III and V is VIII. Only with the cipherization of a number system, such as in the alphabetic notation, and in our own, did it become necessary to memorize that γ and ϵ is η, or that 3 and 5 is 8. However, no thinking person regards the memorization which this entails – at least in our own system – as indicating a retrogression or even as constituting a serious drawback, for it is eminently justified by the increased facility which such a type of numeration affords. This preliminary training which a 'learned' system requires, but which is unnecessary in more primitive 'folk' notations, undoubtedly accounts in large part for the thoughtless criticism of the alphabetic numeration. Scholars hesitate to make the modest initial effort necessary to master a system which they have already pre-judged as difficult to learn and clumsy to use.

In comparing the Ionian system with our own, one is struck immediately by the fact that almost three times as many distinct symbols are involved in the former as in the latter. Nevertheless, one should bear in mind that the twenty-seven letters used as numbers by the Alexandrian Greeks had already been memorized in sequence (with the exception perhaps of three) for the purposes of language, so that in this respect possibly less effort was required of them than is demanded of us in memorizing ten numerical symbols in serial order above and beyond an ordered alphabet of twenty-six letters – thirty-six in all. It appears that to associate numerical values with the letters of the Greek alphabet – that is, to know that ι is the tenth letter in the enlarged scheme, κ the eleventh, ρ the nineteenth, and so on – may well have constituted for them a slighter tax on the memory than does the introduction in our case of ten quite unfamiliar new symbols in order.

More seriously, however, Cantor, Cajori, Karpinski, Menninger[12], and others have proposed an objection which on the face of things appears to be plausible. They point out that tables of addition and multiplication in the Ionian notation are far more elaborate than are ours. The sum of β and γ, they say, affords no analogy to the sum of κ and λ or of σ and τ, such as exists in our system between 2 plus 3 and 20 plus 30 or 200 plus 300; and that β times γ equals ς, γ times κ equals ξ, κ times λ equals χ, and so on, are entirely separate things to remember. Closer analysis shows that this argument too has been overworked. Tables of addition and multiplication are memorized in terms of words and ideas rather than symbols, and the Greek words and concepts for thirty and fifty, for example, would betray the same connection with three and five as do ours. Moreover, the Greeks early introduced the idea of a pythmen or root or order number of each literal symbol within its own rank or versus. That is the pythmen or analogue of α and ι and ρ was one in each case; of β and κ and σ it was two; and so on. The problem of adding σ and τ thus became one of adding the pythmenes two and three to obtain five and then of knowing that this was the pythmen of ϕ; and similarly for products. The numerical association with alphabetic symbols having once been learned, the system presented the same analogies as are to be found in ours, as Delambre, Tannery, and Heath[13] have clearly indicated.

Several further incidental objections have on occasion been raised against the alphabetic notation. One is that the system encouraged gematria, that is, numerology and divination of the type to which Hippolytus animadverted in his *Refutation of All Heresies*, in which words and names are associated through the pythmenes of the letters with the supposed mystic properties of numbers. This argument cannot, of course, be taken seriously. Would one discard the mathematics of Pythagoras or the astronomy of Kepler because of the gross superstition with which these two were so unnecessarily and yet so closely linked?

Another argument holds that the preempting of the letters of the alphabet for purposes of numeration precluded their use as symbols for unknown quantities in algebra, thus seriously delaying progress in this subject. This point is easily seen to be groundless. Would it be more difficult for Greeks to invent new symbols for unknowns than it was for some nameless person or people to devise the ten symbols which we have adopted as digits, thus freeing the letters for algebra? The question is answered by the simple fact that Diophantus did indeed invent and use symbols for unknown quantities. That he went no further in this direction can therefore be attributed to the fact that he felt no need of others, rather than that he lacked the imagination necessary to devise suitable symbols.

The possible further argument that the numerical meanings of letters would be confused with the ordinary use in language is answered through the fact that in the rare cases where such ambiguity arose in the context, letters used numerically were easily and clearly distinguished from others by a line or accent placed over them.

The arguments against the Ionian system of numeration are thus seen on closer inspection to vanish, except in the case of the necessary mental association of numerical values with the twenty-seven letters. This is truly a slight disadvantage at worst, and one which no longer operates after the preliminary effort of memorization has been made. But even this trivial point might easily have been removed and our present system achieved if the Greeks had taken one more step – a step which Delambre[14] over a century ago recognized as but a short one. We have seen that for the first nine multiples of one thousand the Alexandrians adopted the very same symbols as for the first nine natural numbers. In this case, a distinguishing stroke was generally added to the letter, as already indicated, but occasionally this was omitted where the positional value was clear, as in $\theta\tau\mu\theta$ or 9349. Here we have definite recognition and use of a limited principle of local value, which, if adopted also for the tens and hundreds, would have made the Ionian system of numeration precisely like our own. Nor were the Greeks, at least of the later Alexandrian period, unfamiliar with the use of a symbol for zero which such an extension called for. Ptolemy consistently used the letter o to indicate a blank position in his sexagesimal notation. Such a practice may well have been introduced much earlier among the Greeks, as it certainly had been among the Babylonians. Nevertheless, such a symbol was not introduced into the Ionian decimal notation and hence complete use of the principle of local value was not made. Mathematicians and historians have expressed amazement at this fact and many have seen in it an endemic weakness of the Greek mind and culture which was to be contrasted with the skill and intuition of the Hindus. Halsted in this connection said euphemistically, if mistakenly:

> The importance of the creation of the zero mark can never be exaggerated. This giving to airy nothing, not merely a local habitation and name, a picture, a symbol, but helpful power, is the characteristic of the Hindu race whence it sprang. It is like coining the *Nirvana* into dynamos.

In view of the above analysis of Ionian numeration it would appear that an equally plausible explanation, far more elementary and immediate, may be offered in defense of the Greek view. It is entirely possible that the Greeks were aware that, through the use of the principle of local value and of a symbol for an empty position, one could dispense with all but the first nine alphabetic numerals, but that they felt that there was little to be gained through the change. After all, would such a form as δoo be superior in any way to v for 400, or would $\epsilon oooooo$ be a significant improvement over ϕM? Let him raise his voice in criticism who has never either determined incorrectly the number of zeros in a product or made an error through placing a digit in the wrong column. As Tannery remarked, the alphabetic notation offers certain definitive advantages, and the Greeks may well have regarded these as justifying its retention.

In evaluating principles of notation one must keep clearly in mind that small differences in the amount of preliminary training are of far less consequence than the ultimate facility which is afforded. A definitive test of the relative advantages of different systems of numeration would be possible only for those who have familiarized themselves equally thoroughly with each of the notations in question. Such tests, conducted with all necessary statistical precautions, would be very desirable but would be exceedingly difficult to carry out impartially. Few subjects would be found willing to give up the time necessary to master a second system with comparable thoroughness. However, the author of this paper has little doubt that in such a test the Hindu-Arabic system would be found superior to the Ionian. He is equally confident that the arguments here presented would be confirmed by a result showing both of these notations definitely superior to any non-ciphered system, whether positional or not.

This rehabilitation of Greek numeration may perhaps best be closed by a frank admission of the Achilles heel which the badly-aimed darts of critics seem to have missed. Archimedes in the *Psammites* and Apollonius in the lost work on *Pythmenes* presented simple schemes for representing exceedingly large numbers in terms of octads and tetrads. Archimedes found as an upper bound for the number of grains of sand in the universe a thousand myriads of the eighth order, a number which we should write as 10^{64}. Pappus, following the scheme of Apollonius, multiplied together numbers the product of which he expressed as ϛϛ myriads of 13th order and $τξη$ myriads of 12th order and $δω$ myriads of 11th order, or 196,036,848 followed by 46 zeros. There appears to be no corresponding study among the Greeks of notations for very small numbers or fractions. In this connection they adopted in general the devices used by the Egyptians and Babylonians. The former people avoided operations on general rational fractions through the use of elaborate tables expressing proper rational fractions as sums of unit fractions – that is, fractions in which the numerator is unity. The Babylonian treatment of fractions was suggested by their positional notation for integers; they allowed their symbols indifferently to represent not only integral multiples of powers of sixty but also the quotients of integers divided by powers of sixty. Egyptian influence among the Greeks was sufficiently strong to cause the latter to make extensive use of unit fractions[17], but the greater adaptability to computation of the Babylonian scheme led later Alexandrian astronomers to adopt sexagesimal fractions in which, however, alphabetic symbolism was substituted for the cuneiform. Here too the power of the Ionian notation is apparent when one compares the Ptolemaic form of such a number as the chord of 120° in a circle of radius 60 – $ργ$ $νε'$ $κγ''$ – with the corresponding Babylonian form. However, the incongruity of adopting sexagesimal fractions to complement a decimal numeration for integers is all too apparent in our own common measure for angles and in our units of time. Homogeneity, or strict adherence to the modulus of a given scale of notation, is one of the chief criteria of a satisfactory system of numeration or of weights and measures. In complete disregard of this fact our decimal society continues to tolerate such misfits as the octesimal fractions of the New York Stock Exchange.

The Greeks had at hand a notation which could easily have been extended to include decimal fractions[16] through the simple expedient of making further use of the diacritical marks which characterized their decimal system of integers and their sexagesimal system of fractions. For example, they might have employed the mirror image property of the 'Arabic' numbers before and after the decimal point by writing 1234.4321 as $,ασλδ\ \overline{δλσα}'$ or in some equivalent form not significantly less concise and convenient than our own and far more satisfactory than the early decimal notations of Stevin. It is to be remarked that such an extension called for no new devices beyond those of independent representation and the cash-register principle. The decimal point, a symbol for zero, and our complete positional notation can no more be regarded as *sine qua non* of a convenient fractional representation than of a simple numeration for integers. The contribution of Stevin might easily have been made before the introduction of the Hindu-Arabic notation, for it consisted simply in making clear the great advisability of homogeneity – of employing for fractions the same radix and notation as that used for integers. In making this extension Stevin at first felt it necessary to use indices for tenths, hundredths, thousandths, and so on, ingenuously betraying perhaps that place value was almost an after-thought, with tabular cipherization constituting the basic principle.

The shortcoming of Greek fractional notation indicated above is more particularly a deficiency in practical application than a defect in the general principles of Ionian numeration, and hence it does not significantly affect the argument of this paper. What has been said above in the comparison of iterative, positional and ciphered systems of integral notation will hold with equal force for fractions. Is it pertinent to observe in this connection that our own system is habitually known as the Hindu-Arabic, in complete disregard of the fact that neither among the Hindus nor among the Arabs did the notation include general decimal fractions. If one accuses the Greeks of myopia in this direction the same must be said of the Hindus and the Arabs as well. However, in Greek mathematics there were factors not found in the case of the other civilizations which may have operated to obscure the desirability of extending the system to cover fractions. These are the very same factors which limited Greek accomplishments in the direction of the calculus — an insistence on rigor which magnified the difficulties of the ratio of incommensurables and led to the rejection of the mathematical infinity. The word 'number' among the Greeks was taken to mean a collection of units, so that what we regard as a single rational number was looked upon by them as a ratio of two natural numbers. This may explain why common or vulgar fractions do not appear more frequently in Greek arithmetic. Moreover, the decimal representations of rational fractions are more often than not non-terminating. Inasmuch as the infinite had been banned from rigorous Greek thought, theoretical arithmetic would find little use for such representation. On the other hand, the field in which logistic or practical calculation found greatest need for accurate fractional approximations was astronomy, and here Babylonian tradition had imposed the sexagesimal notation which even in our own day has not been entirely superseded.

The main thesis presented here has been that the Greco-Egyptian cash-register cipherization — rather than the zero and positional principle of the Babylonians — represented the most important step in the development of numerational notations. There are two considerations which come to mind in this connection. One of these concerns that seductive quasi-Marxian theory which holds that scientific progress is dictated primarily by economic and social conditions and that mathematics is 'the mirror of civilization'. The contrast between the practicable numeration of our own profit-minded civilization and the supposedly ill-contrived alphabetic notation of the philosophical leisure class of Greece appeared to present a strikingly apt illustration of this point of view. Crowther[17] has well expressed this in saying that Greek geometrical method 'satisfied the intellectual demands of a leisured class. Simultaneously, the social status of arithmetical calculation, which had been relegated to slaves who worked with the abacus, was depressed further [by the Ionian system of numeration!] owing to the widening class differences'. Hogben[18] and others would explain the putative Greek deficiencies in calculation on the basis of social impulses and the fact that they inherited a 'culture which forced them to use a number script evolved before the need for elaborate calculation with large numbers was keenly felt'. However, when one recognizes that the Ionian system of notation was in reality an eminently practical one scarcely less well adapted to computations than that of our present culture and far superior to that of any other civilization in antiquity, such arguments serve only as effective boomerangs.

Another consideration which comes to mind is the bearing, if any, which this defense of Ionian numeration may have on that tantalizing question of the origin of the so-called Hindu-Arabic numerals[19]. It does not, indeed, present any new historical evidence in this connection, but it may serve to place the problem in better perspective. It has been seen that of all the ancient schemes of numerical

representation the Egyptian hieratic and the Ionian alphabetic systems come
closest in general plan to the so-called Hindu-Arabic type. The latter would, in fact,
follow as a simple reasonable consequence from the former through the full
application to it of principles – those of zero and local value – which even then were
known to the Babylonians and were to a limited extent adopted by the Alexandrian
Greeks. It is tempting to come to the conclusion that the general scheme of the
Hindu-Arabic system – although not necessarily the forms of their numerals – was
an historical as well as logical corollary of the Ionian and Babylonian. However,
history is not always logical and it is possible that the Hindus came upon their
system quite independently of any such development as that indicated above.
Moreover, attempts to link with Egyptian hieratic notation or with the letters of the
Greek alphabet the particular forms either of our numerals or of those used by the
Hindus have not been successful. It is possible – however difficult it may be to
believe – that the scope of the recognized influence of Greek and Babylonian
culture in India, during precisely the period when the Ionian notation was being
adopted in the Greek world and astronomy was flourishing in the Seleucid,
somehow did not include numeration. It is also within the realm of possibility that
the alphabetic numerals which entered India at that period and continued to be
used until the time of Aryabhata[20] may have had no connection with those of
Greece, whereas the simultaneous use of these in the case of the Goths, Hebrews,
Syrians, Aramaeans, Persians, and Arabs in all probability did ...

Possibly history never will be able satisfactorily to unravel the tangled bits of
evidence in this situation; but one thing is clear. Whoever invented our notation
had no need to devise new principles, for the two fundamental elements of our
system were known at least four thousand years ago. One of these – the positional
principle – was introduced by the Babylonian civilization. The other – cash-register
cipherization – was suggested by the Egyptians and effectively utilized by the
Alexandrian Greeks. It remained only to incorporate these two principles into one
system, decimal or otherwise, which should take full advantage of each. Cipherization afforded a conciseness which facilitated the writing and reading of numbers
and which lent itself readily to tabular computations; local value supplemented this
by limiting greatly the number of necessary ciphers, thus making mastery of the
system relatively easy. Any exposition of the principles of numeration which fails to
mention both of these aspects is a misrepresentation of the true situation. In the
past it has been customary to afford full credit to local value, without mention of
cipherization. The gross injustice of such a view is quite apparent from a review of
the Babylonian and Ionian systems and from a comparison of these with our
present notation. The fact that in structure the ciphered Ionian notation comes far
closer to the so-called Hindu-Arabic system than does the positional Babylonian
numeration indicates that one might with greater justice regard cipherization as the
basic principle, with local value playing a subordinate role in tempering the
multiplicity of characters which are called for in a ciphered system.

Boyer's views are presented at length in his own words because they provide an
interesting and illuminating interpretation of the relative importance of *place, value*
and *cipherisation* in the development of an efficient fully cipherised place-value
system of writing numbers. It is, however, not the only view; indeed, as the number
of authors cited for criticism in the extract suggests, it is if anything a minority view.
Later, the development of our fully cipherised decimal place-value system of
numerals will be discussed and it should prove useful to refer again, when studying
these sections, to some of the points raised in Boyer's paper (e.g. see particularly
pp. 95–6 above).

The principle of using letters of the alphabet to represent numerals is of Semitic
origin, though their allocation in alphabetic order to 1–9, 10–90 and 100–900 is
certainly due to the Greeks. This formula of allocation was subsequently adopted

by Hebrew writers, who used the 22 letters of the Hebrew alphabet together with
seven special end-forms of letters to obtain the necessary 27 symbols (see Figure
56). When the Greek alphabet was used as a model for others, such as the Coptic,
Gothic and Cyrillic alphabets, the alphabetical numeral system was also adopted.
There is, however, one interesting variation, the *Glagolitic* alphabetical numerals
(see Figure 57). The basic principles are the same, but the order in which the letters
are allocated to numerals has a number of anomalies. The Glagolitic alphabet was
invented (probably by St Cyril) in order to provide the written Christian Gospels in
the Slav language. Scholars differ as to the origin of the forms of the Glagolitic
literals, though the consensus is that, like the Cyrillic, they derive from Greek

1–9	א ב ג ד ה ו ז ח ט
10–90	י כ ל מ נ ס ע פ צ
100–400	ק ר שׁ ת
500–900	צ (final) ף (final פ) ן (final נ) ם(final מ) ך (final כ)

FIGURE 56   Hebrew numerals including end-forms

Glago-litic	Nume-rical value	Cyrillic	Nume-rical value	Glago-litic	Nume-rical value	Cyrillic	Nume-rical value
✢	I	ⰀА	I	Ⱁ	700	Ѡ	800
Ⰵ	2	Б	—	Ꙋ	800	ЏЩ	—
Ⰲ	3	в	2	Ⱛ	900	Ц	900
Ⰳ	4	г	3	Ⱌ	1,000	Ч	90
Ⰰ	5	Д	4	Ш	—	Ш	—
Ⰵ	6	е	5	Ⱂ	—	҄һ	—
Ⰶ	7	ж	—	Ⱂ	—	ь	—
Ⰷ	8	Ѕ, ꙃ	6	ⰒⰀ	—	ꙑ	—
Ⰸ	9	Ꙁ, ꙁ	7	Ⰴ	—	ѣ	—
Ⱑ Ⱑ	10	I (ї)	10	Ⰴ	—	ꙗ	—
Ⰸ	20	И	8	—	—	ІЄ	—
Ⰿ	30	(һ)	—	ⱂ	—	ю	—
Ⰽ	40	к	20	Ⰵ	—	Ѧ, ѧ	900
Ⰾ	50	л	30	—	—	—	—
Ⰻ	60	м	40	ꙗ	—	Ѫ	—
Ⱃ	70	н	50	—	—	—	—
Ⱀ	80	о	70	ꙗ	—	ІѦ	—
Ⱆ	90	п	80	—	—	—	—
ь	100	ρ	100	ꙗ	—	ІѪ	—
Ⱂ	200	с	200	—	—	—	—
ⱜ	300	Т, Ш	300	—	—	Ѯ	60
Ⱎ	400	оу, Ѹ	400	—	—	Ѱ	700
Ⱇ	500	ф. ҩ	500	ⱚ	—	Ѵ, ѵ	400
Ⱁ	—	Ѳ	9	—	—	—	—
ь	600	χ	600	—	—	—	—

FIGURE 57   Glagolitic and Cyrillic alphabets and allocation to numerals

forms. However, this does not explain why, for example, the Cyrillic form for *k* represents 20 whilst its Glagolitic equivalent represents 40^{21}, an anomaly which is still the subject of scholarly dispute.

Roman numerals

The Roman numerals consist of the seven symbols:

$$I = 1$$
$$V = 5$$
$$X = 10$$
$$L = 50$$
$$C = 100$$
$$D = 500$$
$$M = 1000$$

In the earlier discussion of the origin of these symbols in relation to number words and tally marks it was seen that in old inscriptions an arrow-shaped symbol instead of L (= 50) is sometimes encountered. The number XXCIIII (= 84) in line 4 of the inscription in Figure 58 (translated on p. 50) reveals the repetitive, additive and

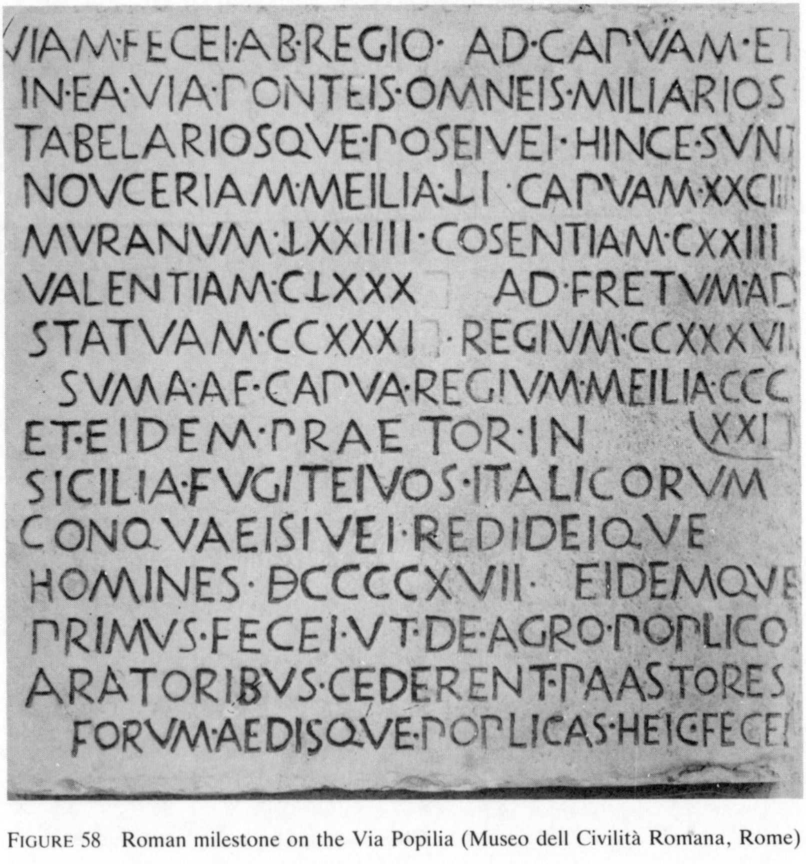

FIGURE 58   Roman milestone on the Via Popilia (Museo dell Civiltà Romana, Rome)

subtractive principles present in the Roman system. Also, 4 is written as IIII and not as IV; the latter form dates from the late Middle Ages and is not found in classical sources.

The symbols I, V, X, as shown earlier, are not the initial letters of the number words *unus, quinque* and *decem* but are almost certainly derived from notches on tally sticks. The symbols C and M are the first letters of *centum* and *mille*, though on older inscriptions we also find an alternative symbol ⊂ɸ = 1000, the right half of which suggests the symbol D = 500.

For 10000 and 100000 the old inscriptions sometimes include the symbols ⊂⊂ɸ and ⊂⊂⊂ɸɸ respectively. The latter is found, for example, on the inscription from the *Columna rostrata* where it is repeated thirty-two times (see Figure 59). Halving these signs yields the symbols ɸ = 5000 and ɸɸ = 50000. Variants of these

FIGURE 59 *Columna rostrata* inscription (Fototeca, Unione, Rome)

can be found on Roman coins from the first century BC. From the *Columna rostrata* inscription it can be concluded that no sign for one million existed at that time. This is confirmed by Pliny (first century AD), who writes in the *Historia Naturalis*:

> The ancient Romans had no number higher than a hundred thousand; hence, even today this number is multiplied so that one says 'ten times a hundred thousand' and the like.

Later on, as millions came into more frequent use, a special symbol was introduced for multiples of 100000, as shown in Figure 60.

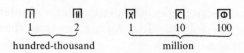

hundred-thousand   million

FIGURE 60   Large Roman numerals

Mayan numerals

The Maya lived (and still live) in Yucatan in Middle America (see Figure 61). Their civilisation reached its peak in what is called the 'Classical Period' between AD 300 (or even earlier) and 900. There are numerous inscriptions from this period, a large number of which consist of dates of events. Thus there is considerable knowledge of the Mayan calendar and of the Mayan method of writing large numbers.

FIGURE 61   The Mayan empire

The Maya had three kinds of calendars. The first was based upon a sacred period of 260 days, the *tzolkin*, which was divided into 20 periods of thirteen days each. The fundamental period of the second calendar was a round year of 265 days called *haab*, which was divided into eighteen divisions of twenty days each and a short division of five days.

Large numbers were used mainly in the *long count*. The fundamental unit in this long count was the *tun* of 360 days. It was divided into eighteen *uinals* of twenty days each. Twenty *tun* were called *katun,* because *ka* or *kal* was the word for 20. Twenty katuns were called *baktun,* because *bak* was the word for 20^2. Thus a baktun was nearly 400 years.

FIGURE 62   Mayan 'named' place-value notation

For each of the units: *baktun, katun, tun, uinal* and day, a special hieroglyph was used, which in most cases had the form of a head with distinguishing attributes. This hieroglyph was accompanied by a numeral indicating how many baktuns, katuns, etc. one had to take. The numerals were made up of bars and dots, each bar meaning 5 and each dot meaning 1. Thus the number 17, for example, was written as ☰ . Often if a number, such as 17 or 12, required only two dots an ornament in the form of an ear was placed between them to fill up the space. Thus the five hieroglyphs shown in Figure 62 (to be read from left to right and from top to bottom) denote elapsed times of

$$
\begin{array}{llll}
9 \ baktun & = 3600 \ tun \\
14 \ katun & = 280 \ tun \\
12 \ tun & = 12 \ \ tun \\
4 \ uinal & = & 80 \ \text{days} \\
17 \ \text{days} & = & 17 \ \text{days} \\
\hline
\text{Sum} & = 3892 \ tun \ \text{and} \ 97 \ \text{days} \\
& = 1\,401\,217 \ \text{days}
\end{array}
$$

Usually the end of such a long time (more than 3800 years) was a recent date such as the founding date of a temple, whereas the beginning was the initial part of a 'long count' beginning in a mythical past.

The Maya also had a special sign for *zero*, which they could insert instead of a numeral. Thus, the three columns in Figure 63 would mean:

$$
\begin{array}{l}
8 \ tun \ 2 \ uinal \ 0 \ \text{days} = 2920 \ \text{days} \\
16 \ tun \ 4 \ uinal \ 0 \ \text{days} = 5840 \ \text{days} \\
9 \ baktun \ 10 \ katun \ 5 \ tun \ 0 \ uinal \ 2 \ days = 3805 \ tun \ \text{and} \ 2 \ \text{days}
\end{array}
$$

Figure 63 shows that the Maya achieved an abstract place-value notation with a zero based on the number 20. This has been considered something of an enigma.

FIGURE 63   Mayan abstract place-value notation containing the oldest zero in the New World

For example, Menninger writes:

Thus we have an enigma of cultural history: an abstract place-value notation with a zero, based on the number 20, occurring in apparent isolation, far away in the New World. A native invention, or an indirect borrowing from India? Or was it even borrowed directly? Since the extinct Mayan culture was at its height during the period from the sixth to the eleventh centuries A.D., China may be eliminated as the intermediary, for the zero sign was first brought from India to China in the middle of the thirteenth century.[22]

Early Hindu numerals

In its final form the Hindu system of writing numerals is fundamentally different from that of the Egyptians, the Ancient Greeks and the Romans. Whereas these possess special number symbols for the individual *ranks* (units, tens, hundreds, thousands, etc.), which are lined up side by side as necessary the Hindus have special symbols for the individual *numbers* from one to nine, whose ranks are indicated by their positions. The Hindu system is to some extent a continuation of the early Chinese system, although it has not been proved that there was a direct influence. Like the Hindu, the Chinese system has individual symbols for the numbers 1 to 9, but the ranks are also written down, thus making the system a *named place-value system*. The Hindu system is a *pure place-value system*. Only a pure place-value system needs a symbol for a missing amount, for a non-existent rank, the *zero*. The only pure place-value system comparable to the Hindu is that of the Sumerian-Babylonian culture but, as shown earlier, for many centuries there was no zero in this system, whilst in the development of numerals in India the zero made a relatively early appearance. Only the Hindus within the context of Indo-European civilisations (and in their wake the Arabs and the Europeans) have consistently used a zero.

The Hindu place-value system has assumed great historical importance because the familiar modern decimal system is directly descended from it via the Arabs. The remainder of this chapter is devoted to the reconstruction of this descent. First, the origins of the decimal system among the Hindus will be discussed, for, as with the system of other cultures (discussed earlier), it is the culmination of a long development, a development which can be reconstructed, partially at least, from existing sources.

The flowering of the so-called *Indus culture* occurred at about the same time as the great river cultures centred on the Nile and on the Tigris and Euphrates rivers (*c.* 3000 BC). In this century extensive ruins have been uncovered around Mohenjo-daro and Harappa (see Figure 64), bearing witness to the fact that large, well-planned and well-constructed towns existed there. From the ruins of public buildings, from watering and drainage canal systems and from paintings on ceramic articles it may be concluded indirectly that a certain amount of mathematical knowledge was one of the elements which made up a highly developed culture. Even a ruler with decimal divisions, made from a shell, has been found. Many inscriptions in the manner of picture-writing have been preserved on articles of daily use and on seals. Unfortunately the language has not yet been deciphered. However some of the seals and inscriptions include signs which seem to be number symbols: vertical strokes and groups of strokes, ranged either side by side or beneath one another (see Figure 65). It appears that at least the numerals 1–13 were indicated in this way; but an unambiguous connection cannot be proved at present. Likewise it cannot yet be said whether or not there were symbols for larger numbers, 20, 30, . . ., 100, etc.

There is a gap of at least 2700 years between the discoveries of the Indus culture and the oldest inscriptions containing number symbols from the Aśoka era. However, the fact that in literary texts from the intervening period very large numbers are mentioned and that in the Aśoka inscriptions number symbols are used as a matter of course does suggest that well before 300 BC a numeral system existed.

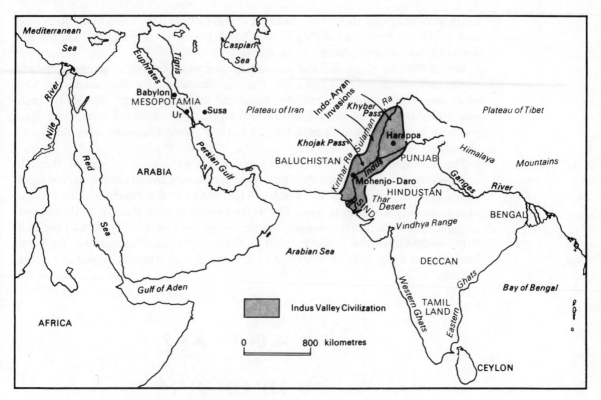

FIGURE 64   Ancient India

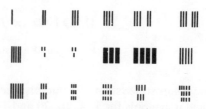

FIGURE 65   Indus culture symbols

From the middle of the second millennium BC Indo-European tribes were making their way from the north-west towards India. They conquered the indigenous population and introduced Sanskrit, the language which remained the learned language of India for thousands of years.

The earliest knowledge of mathematics in India dates from the time when the sacred religious-philosophical books of the 'Veda' were composed. The *Rules of the Cord (Śulba-sûtra)* belong to the latter part of this era and contain predominantly geometrical constructions and calculations. In these, and in still older documents of a religious character, number words for very large numbers are often found. They show that for ages the Hindu number sequence had been decimal. In Sanskrit there are number words for 1–9, for 10, 100 and the further powers of 10 (see p.72). One of the Hindu characteristics is that in the terminology for the powers of 10 they did not, like the Greeks, stop at 10^4 or, like the Romans, at 10^3; they already had signs for powers of ten up to 10^{53} by the third century BC. It does seem likely therefore that the Hindu system of writing numerals was based on the number 10 in the era for which no written evidence is available.

Before the rise of the decimal place-value system, there were various alphabets in the area which today is India and thus various ways of writing numerals. The most important alphabets are the *Kharosthî* and the *Brâhmî* alphabets, both established by decrees of the Emperor Aśoka (*c.* 250 BC). In this section we shall see how numerals were represented in both of these (basically very different) alphabets.

The Kharosthî script which is written from right to left, spread through north-western India between the fifth century BC and the third century AD, above all in the province of Gandhara (eastern Afghanistan and northern Punjab). Gandhara had been the eastern extremity of the Persian empire since the sixth century BC, and became after its conquest by Alexander the Great (327–325 BC) the link between the Greek and Indian cultural traditions. Kharosthî is a typical scribe's and businessman's alphabet; it developed from the Syrian-Aramaic alphabet via the Persian.

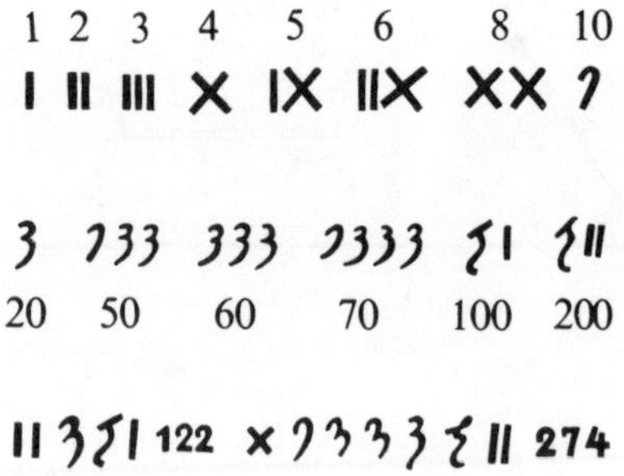

1	2	3	4	5	6	8	10
I	II	III	X	IX	IIX	XX	?

20	50	60	70	100	200

II ?| 122   X ??? ?II 274

FIGURE 66   Kharosthî numerals

The Kharosthî system of numerals (see Figure 66) is not built up uniformly; it has particular symbols for 1, 4, 10, 20 and 100 (as shown). The numerals from 1 to 8, and almost certainly 9 (this numeral not being found in the texts preserved) were composed according to the additive principle from the figures for 1 and 4. It is not clear why the four should be chosen; this was originally represented by four vertical strokes, later substituted by a cross. The method used here of writing the numerals from 4 to 8 is not found in earlier Aramaic texts. The symbol for 20 consists of two symbols for 10 written with one stroke. Twenty groups, which have come down to us from number speech (see pp. 30–33) occur very rarely in written form. The method of representing tens with the aid of symbols for 10 and 20 is the same as that used by the early Phoenicians and Aramaics. The hundreds were written in a very different way from the tens, namely by multiplication. Next to the symbol for 100 (not used independently) were placed the figures indicating the number of hundreds – the same kind of system as that found in China (see pp. 84–87).

The Brâhmî alphabet is historically more important than the Kharosthî and can be regarded as the mother of all Hindu alphabets, and in particular of the one most widely disseminated today, (Deva-)Nagari. Since the time of Aśoka (third century

BC), who united almost the whole of India under his rule, Brâhmî inscriptions have been found all over the subcontinent. The origin of the Brâhmî alphabet, which in contrast to the Kharosthî is written from left to right, is disputed: according to Hindu opinion it is native, whereas other historians consider it to have stemmed from the North Semitic (perhaps Phoenician) group of alphabets.

Brâhmî numerals were used for over 2000 years, and longest of all in Sri Lanka (Ceylon), where they arrived at the same time as Buddhism and were in use up to the nineteenth century. Unfortunately, it is not possible to give the precise forms of the original Brâhmî numerals; Figure 67 shows symbols from an early text of the third century BC. A little later those symbols appear which for nearly 1000 years occur in essentially the same form on temple and cliff walls, on copper plates, and in other written representations. The oldest document to contain most of these numeral forms comes, it would appear, from the second or first century BC. It was found in a cave on the hill of Nânâghât, 75 miles from Poona.

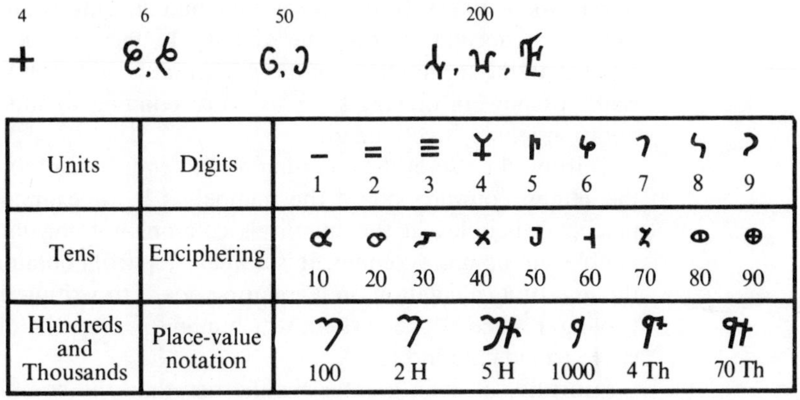

				200
4	6	50		

Units	Digits	−	=	≡	Ƴ	ſ	ϥ	7	ϟ	ϡ
		1	2	3	4	5	6	7	8	9
Tens	Enciphering	α	๑	ๅ	✗	ﾉ	˧	✗	⊙	⊕
		10	20	30	40	50	60	70	80	90
Hundreds and Thousands	Place-value notation	ワ 100	ワ 2 H	ꓘ 5 H	᧚ 1000	᧚ 4 Th	᧚ 70 Th

FIGURE 67   Brâhmî numerals and old Brâhmî numerals (above), 4, 6, 50, 200

Since the main interest in Brâhmî figures here is their importance in relation to our own numerals, discussion of the theories of the origins of the various symbols would not be relevant.[23]

One characteristic typical of the Brâhmî system of numeral writing is that the units are no longer built up by the law of accumulation and grouping that we discern in the other number systems, but rather each unit receives one individual symbol, one 'figure'. The existence of special symbols for the figures 1–9 was a typical and important characteristic of Hindu arithmetic and provided the prerequisite for the rise of the decimal place-value system. This numbering is not, however, restricted to the units; the tens are also written with individual symbols. This latter must be regarded as an obstacle on the way towards a consistent place-value system.

From the number 100 onwards, numbers are represented according to another principle: there are only two new symbols (for 100 and 1000); it is easy to see that further hundreds and thousands could be indicated by combinations of these with the symbols for the units. That is to say: instead of numbering in the same way as the units and tens, we have here another *named place-value system*.

The introduction of place-value in India

In India, probably in the sixth century AD, the system of Brâhmî numerals was transformed into a genuine place-value system. Did this transformation take place without external influence, or was it influenced by ideas from outside India? The answer to this is not known at present, but several circumstances can be enumerated which were favourable to the development of a decimal place-value system in India.

First, the Hindu counting system was purely decimal, and distinct symbols for the numbers 1 to 9 existed already before the invention of the place-value notation.

Secondly, a notation for high powers of 10 existed at an early stage. Prince Gautama Bodhisattva knew (according to the *Lalitavistara*) number words for 10^7, 10^9, ... up to 10^{53} and he said that there were eight more such sequences beyond this. Although such numbers were probably never used for calculations the text does indicate that Hindu mathematicians were aware of the fact that the sequence of powers of 10 can be continued indefinitely. This awareness is an important step towards a decimal place-value system. In fact, if one were to express a large number in the terminology of the *Lalitavistara* and then leave out the names of the ranks of powers of ten, symbols corresponding to modern decimal place-value notation would be obtained.

Thirdly, the use of the counting board tends to lead to a place-value system. On the Hindu counting board the number 702, for example, would be laid out by placing 7 pebbles in the hundreds column, nothing in the tens column, and 2 pebbles in the units column. It is a small step from obtaining a result such as 702 at the end of a calculation on a counting board to writing down symbols for 7 and 2 leaving a space between them which might perhaps be filled with a dot or a small circle to indicate 'no tens'.

Fourthly, of even greater significance than the counting board seems to have been the astronomical activity in India during the period from AD 490 to 630. The famous astronomer Âryabhaṭa, whose chief work the *Âryabhaṭîya* is extant, lived at the beginning of this period. Equally famous was the astronomer Brâhmagupta, who lived at the end of the period. His chief work was the *Brâhma-sphuta-siddhânta* (*Improved Astronomical Textbook of Brâhma*). Between the times of these two astronomers there lived (about AD 550) the astrologer Varâha Mihira, who wrote a book, the *Pañcasiddhântikâ* (*panca* = 5), in which the contents of five siddhântas (astronomical textbooks) were summarised. Three of these textbooks used trigonometrical methods and were based on pre-Ptolemaic Greek astronomy, while the other two were less accurate and used Babylonian methods.[24] In all five siddhântas the Babylonian division of the zodiac (see p. 83) was used. This astronomical activity had important consequences. Astronomers like Âryabhaṭa and Brâhmagupta worked with large periods such as the 'Great Year' (*Mayâyuga*) of 4 320 000 years, and therefore had to work with large numbers. For this purpose, an efficient system of writing numbers was highly desirable. Also the Hindu astronomers had before them the Babylonian place-value system, with base 60. They could very well invent, by analogy, a place-value system in their own counting base, namely 10. The Hindu astronomers were also familiar with Greek astronomy and its notation. In Ptolemy's tables a small circle was used to indicate the absence of degrees, minutes or seconds. (It is probable that the small circle was the first letter of *ouden* = nothing). It is quite possible that Ptolemy's predecessors used this small circle and that the Hindu astronomers knew of this and imitated it in their decimal system.

From all these considerations it can be seen that in India in the sixth century AD the conditions for the development of a decimal place-value system were favourable. We shall investigate the further steps in this development.

Âryabhaṭa is the earliest Hindu astronomer whose date we can establish with complete certainty. In his *Âryabhaṭîya* he says[25] that he reached the age of 23 years in the year 499. Of course, he did not use the Christian era; he just said that he was 23 years when 3600 years of the Kaliyuga had passed. He invented a notation for writing numbers which was not positional, but came very near to it. In this notation, numerals were written as syllables, as follows: syllables containing the vowel *a* denoted units and tens, syllables with *i* hundreds and thousands, syllables with *u* multiples of 10^4 and 10^5 and so on. In Sanskrit there are nine vowels:

$$a \ i \ u \ r̥ \ l̥ \ e \ ai \ o \ au$$

so the vowels can be used to denote powers of 10 up to 10^{17}. If higher powers are needed, says Âryabhaṭa, one can start again with *a, i, u*, etc. The consonants are used to indicate how many units, tens, etc. one has to take. In the Sanskrit alphabet there are 25 *varga* or classified letters from *k* to *m*, and eight *avarga* letters from *y* to *h*. The 25 *varga* letters are used by Âryabhaṭa to denote the numbers from 1 to 25, the eight *avarga* letters to denote the numbers 30, 40, etc. up to 100. In writing a number one has to start with the lower ranks (units, tens, etc.). A zero is not needed in this system; missing places are just omitted.

For example, 57 753 336 (the number of revolutions of the moon in a Great Year according to the *Âryabhaṭîya*) in Âryabhaṭa's syllable notation can be arranged in pairs:

57, 75, 33, 36

If one considers the pairs in opposite order, starting with *thirty six*, or rather *six and thirty*, because one has to take the units first, then for 6 and 30 one has to take the consonants *c* and *y* and to combine them with the vowel *a*, thus obtaining *caya*. For the next pair, 33, one has to use the vowel, *i*, which gives *giyi*. The next pair, 75, gives *ṅuśu*. Taking the last pair, one combines the consonants *ch* = 7 and *l* = 50 with the vowel *r̥*, thus obtaining *chlr̥*, and the result is

57 753 336   *cayagiyiṅuśuchlr̥*

Âryabhaṭa's system is not a place-value system, and from the fact that such a complicated system was invented it may be concluded that in his day a place-value system in India did not exist. Datta and Singh comment as follows:

One advantage of this notation is that it gives very brief chronograms. This advantage is, however, more than counterbalanced by two very serious defects. The first of these is that most of the letter chronograms formed according to the system are very difficult to pronounce. In fact, some of these are so complicated that they cannot be pronounced at all. The second defect is that the system does not allow any great variety in the letter chronograms, as other systems do.[26]

These defects of difficulty in pronunciation and lack of variety were rectified by a pupil of Âryabhaṭa, known to us as Bhâskara I. In his alphabetic notation:

1 is denoted by the letters   *k, ṭ, p, y*
2 is denoted by   *kh, ṭh, ph, r*
3 is denoted by   *g, ḍ, b, l,*
4 is denoted by   *gh, ḍh, bh, v,*
5 is denoted by   *ṅ, ṇ, m, ś,*
6 is denoted by   *c, t, ṣ,*
7 is denoted by   *ch, th, s,*
8 is denoted by   *j, d, h,*
9 is denoted by   *jh, dh,*
0 is denoted by   *ñ, n,* and vowels *standing by themselves.*

For example, the number 644 was written as

bha–va–ti
4   4   6

This is a decimal place-value system; it has a zero, and the powers of 10 are distinguished only by the places they occupy. So we may infer that the first decimal place-value system was invented in India between the time of Âryabhaṭa and that of his pupil Bhâskara I, that is during the period AD 499–522.

There were several later variants of this system[27] all known under the general name of *Kaṭapayâdi systems.* There was, however, another system which was more popular than the alphabetic systems. In this system, individual ciphers were represented by words.

To write a table of sines[28] or a page of a cricket score-book in verse form, is very difficult. This is because number words are too fixed – they have virtually no synonyms. So Hindu astronomers (who wrote their works in verse form) substituted different words for numbers. Instead of 'one' they might write *śaśi* ('moon') because there is only one moon. For 'two' they might use 'eyes' or 'arms' or 'wings'. For 'three' they might use 'fire' (since there were three fires in their mythology) or 'brothers' (because the god Rama had three brothers), and so on. For 'five' they could write either 'senses' or 'arrows' (because the god of love, Kamadeva, had five arrows). It is just as if we were to write 'graces' for 'three' or 'muses' for 'nine'.

As in the Âryabhaṭa and Kaṭapayâdi systems, these *poetic numbers* began with the units, followed by the tens, etc. Thus 867 might be written as:

giri–rasa–vazu (mountains–smells–gods)
7   6   8

A word meaning 'gap' or 'hole' was inserted to indicate an empty place. For example 1021 could be written as:

śaśi–paksa–kha–eka (moon–wings–hole–one)
1   2   0   1

In this manner the pupils of the Hindu astronomers learned an entire table of sines by heart in verse form. The most ancient work known to us in which such a table of sines appears in verse is the *Sûrya-Siddhânta.* An early edition, called the *Old Sûrya-Siddhânta,* already existed by AD 550 when Varâha Mihira wrote his compendium *Panča-siddhântikâ*. (A later, revised edition of this Siddhânta remained a standard work on astronomy in India until quite recently.) The date of composition of the *Old Sûrya-Siddhânta* is given by Datta and Singh, though without giving their reasons, as about AD 300 but such an early date is highly improbable because the *Old Sûrya-Siddhânta*, as we know it from the excerpts of Varâha Mihira, is based on the 'midnight system' of Âryabhaṭa. Also the longitudes of the planets, calculated according to the *Old Sûrya-Siddhânta*, are fairly accurate for AD 500 but *not* for AD 300. For these reasons one may suppose that the *Old Sûrya-Siddhânta* was composed after Aryabhaṭa's day. On the other hand, by AD 550, the *Old Sûrya-Siddhânta* existed already, because Varâha Mihira made excerpts from it. The conclusion is therefore that the *Old Sûrya-Siddhânta* was probably composed about AD 530. In any case, no matter whether this conclusion is correct or not, the word-number method with its zero and its decimal place-value system must have existed by about AD 530.

Word numbers were used quite frequently in the *Agni-Purâṇa*, a religious compendium ascribed by Pargiter[29] to the first centuries of the Christian era. However, other scholars prefer later dates and parts of the compendium may well be later additions. (In Sanskrit literature only a few astronomical treatises, such as the *Âryabhaṭîya*, can be dated accurately.) It would therefore seem that the occurrence of the word-numbers in the *Agni-puraṇa* cannot be used as a valid argument in favour of an early date. It is quite possible that the word numbers were first used in the *Sûrya-Siddhânta*. There is a tradition that the *Sûrya-Siddhânta* was composed by Lata or Latadeva ('the divine Lata'), a follower of Âryabhaṭa. In any case, the composer was a fanciful man! The name 'Sûrya-Siddhânta' means that the contents of the treatise were supposed to have been revealed by the sun god Sûrya himself. The extant text is called the *New Sûrya-Siddhânta*[30] because it differs in some respects from the *Old Sûrya-Siddhânta* summarised by Varâha Mihira. Right at the beginning of the text the story of its revelation by the Sun God is told as follows:

> When but little of the Golden Age was left, a great demon named Maya, being desirous to know that mysterious, supreme, pure, and exalted science, that chief auxiliary of the scripture in its entirety – the cause, namely, of the motion of the heavenly bodies – performed, in propitiation of the Sun, very severe religious austerities.
> Gratified by these austerities, and rendered propitious, the Sun himelf delivered unto that Maya, who besought a boon, the system of the planets.
> The blessed Sun spoke:
> 'Thine intent is known to me; I am gratified by thine austerities; I will give thee the science upon which time is founded, the grand system of the planets. No one is able to endure my brilliancy; for communication I have no leisure; this person, who is a part of me, shall relate to thee the whole.
> Go therefore to Romaka-city, thine own residence; there, undergoing incarnation as a barbarian, owing to a curse of Brahma, I will impart to thee this science.'
> Thus having spoken, the god disappeared, having given directions unto the part of himself. This latter person thus addressed Maya, as he stood bowed forward, his hands suppliantly joined before him: 'Listen with concentrated attention to the ancient and exalted science, which has been spoken, in each successive Age, to the Great Sages by the sun himself'.

Of particular interest in this extract is the reference to 'Romaka-City', which could be any city in the Roman Empire, but quite possibly refers to Alexandria, the mathematical centre of late Greek civilisation. If so the myth does contain this much truth, that the astronomy of the *Sûrya-Siddhânta* was based on Greek sources.

It is quite possible that the man with the poetic imagination to invent this story also invented the method of word numbers. In any case, the poetic word numbers are extremely useful in a treatise containing tables which are meant to be learnt by heart. Numerals cannot easily be incorporated into rhymed verse, but words can.

One thing which the three systems discussed here have in common is that the lowest power of ten (i.e. 10° – units) appear first. It seems that the reversal of the order and the introduction of the ciphers for 1 to 9 to replace syllables or words took place after AD 550, since there is no trace of this reversal in the work of Vahâra Mihira. The next area of discussion will cover the nine numerals used in the later place-value system and will then consider at what point the 'normal' order and the circle (representing zero) came into use.

The use of 'zero'

The most important as well as the most widely used place-value symbols are those belonging to the *Nâgarî* script. Figure 68 shows that the earliest forms of the Nâgarî numerals are very similar to the Brâhmî numerals from 1 to 9 which were in general use before the introduction of the place-value system. This very remarkable situation is found only in India – nowhere else. Our own ciphers as well as those of the Arabs are completely different from the numerals used before the introduction of the place-value system. The explanation is that our ciphers were imported from the Arabs, and those of the Arabs imported from India. In India, however, no ciphers were imported; indeed, there was no need for such importation as the old system of numerals was sufficiently cipherised and its symbols for 1 to 9 were used, together with an additional symbol for zero (a small circle), in the later system.

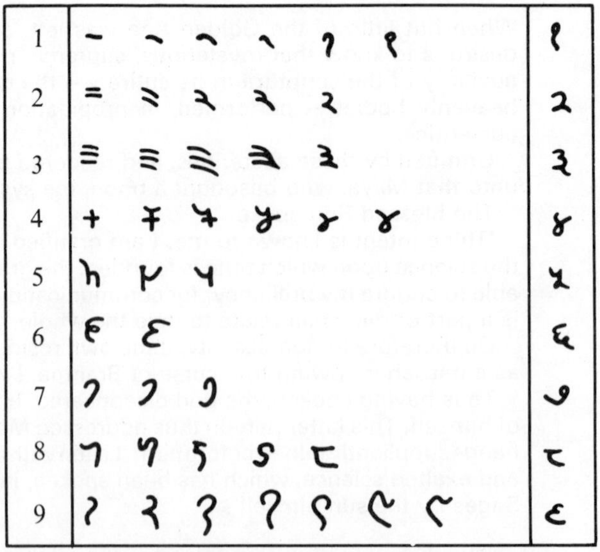

FIGURE 68   Development of Nâgarî numerals

Figure 68 also shows that several of the Nâgarî numerals are very similar to our own. The similarity is striking for the later forms of 2 and 3 and for some earlier forms of 4, 7, 8 and 9. Earlier forms of 2 and 3 show traces of the principle of repetition encountered earlier (e.g. in Egyptian and Babylonian numerals).

A younger contemporary of Vahâra Mihira was Jinabhadra Gaṇi (AD 529–89), author of a work entitled *Bṛhat-kṣetra-samâsa*. In this, the number 224 400 000 000 is expressed as

twenty-two forty-four, eight zeros.

Elsewhere in the same work the simplification of a mixed number

$$241\,960\,\frac{407\,150}{483\,920} = 241\,960\,\frac{40\,715}{48\,392}$$

is described as:

Two hundred thousand forty-one thousand nine hundred and sixty; removing the zeros, the numerator is four-zero-seven-one-five, and the denominator four-eight-three-nine-two.

The *Bṛhat-kṣetra-samâsa* with its 'normal' order of digits, its zero and its place-value notation for the numerator and denominator of a fraction is important for the dating of the invention of the cipherised place-value system. If we suppose that Jinabhadra Gaṇi wrote his book between the ages of 30 and 40, we are led to conclude that the system was invented in India before AD 570. It soon came into general use. On an inscription from Sankheda, the date Saṁvat 346 (AD 595) is written just in the form 346. Two inscriptions from the seventh century, from Belhari (AD 646) and from Kanheri (AD 674)[31], also contain dates written in the decimal place-value notation.

Very probably the earliest symbol for *zero* was a dot and not a small circle. Evidence as to the use of the dot is furnished by the writings of Subandhu, a poet who flourished about the close of the sixth century. In his poem *Vâsavadattâ* he writes:

And at the time of the rising of the moon with its blackness of night, bowing low, as it were, with folded hands under the guise of closing blue lotuses, immediately the stars shone forth, ... like zero dots, ... scattered in the sky ....

A dot for zero also appears on an inscription from Cambodia written in the year 605 of the śaka era (AD 683). The eighth-century Chinese text *Khai-Yuan Chan Ching* (see p. 87) described an Indian numeral system using a dot for zero. On the other hand an inscription from the island of Banka in Indonesia, giving the date 608 Śaka (AD 686), uses a small circle for zero in accordance with later usage. Figure 69 shows an inscription from Gwalior from the year AD 870. In the fourth line from the top (marked by arrows) is the number 270 in place-value notation with the zero written as a small circle. Furthermore in the top line (marked by dots) there is the date 933 (AD 870), and in the fifth line (marked by double dots) the number 187.

Thus it can be seen that both a dot and a small circle were used as early forms for zero, the dot later falling out of use. Evidence for these early forms comes from the

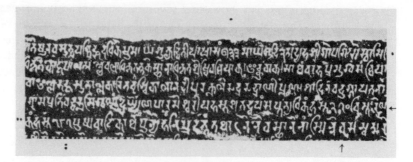

FIGURE 69   Gwalior inscription of the reign of Bhojadeva. In the fourth line from the top (marked by arrows) is the number 270 in place-value notation. In the top line (marked by a dot from above and on the right) is the date 933 (= AD 870). In the fifth line (marked by double dots) is the number 187

eastern side of the subcontinent, Indo-China and south-east Asia. The dot must have been widely used for knowledge of it to have reached China.

Hindu numerals and the Arabs

Knowledge of the Hindu decimal system was early in reaching the West. The first evidence of it is found in the writings of the Syrian scholar Severus Sebocht (AD 662), who lived on the upper reaches of the Euphrates. He refers to significant discoveries in astronomy by the Hindus and states that their method of writing numerals with the aid of nine symbols is deserving of the highest praise. He does not mention the zero, possibly because he did not regard it as a symbol having the same status as 1–9; even al-Khowârîzmî (AD 780–850?) spoke later of nine symbols, although he did make use of the zero.

On its way to the West the Hindu method of writing numerals soon became known to the Arabs, who had established a world empire from the seventh century. In the ten years between Mohammed's flight from Mecca to Medina (AD 622) and his death (AD 632), he had succeeded in uniting under the flag of Islam the Arab tribes formerly wracked with political strife. Less than a hundred years after Mohammed's death an Islamic empire had arisen, which embraced North Africa and Spain as well as the Arabian heartland of the Near East (see Figure 70). But traditions were not destroyed in the subject areas. The native culture persisted; only the Arabic tongue was taken over. The cultural and political centre of Islam was Baghdad. Here foreign texts, above all Greek, Hindu and Persian texts, were collected, translated and adapted. Thus from the eighth century the Arabs enjoyed a unique role as mediators in the field of culture. Many mathematical books from the Greeks and Hindus were translated in Baghdad, travelled west, and after translation into Latin became accessible to Western Europe too.

The Arabs also played an essential part in the dissemination of the numeral system. Originally the Arabs did not possess such a system of their own, but wrote out numbers word by word, even in mathematical texts. Side by side with this they permitted the continued use of the numeral system which had been current in the conquered lands up until then. Even when the use of the Greek language was forbidden (AD 706), Greek alphabetical numerals were retained, though later these

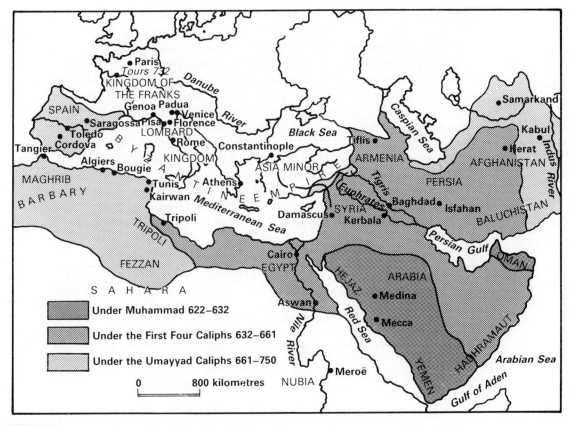

FIGURE 70 The expansion of Islam

were replaced by an analogous system using Arabic letters. There is no evidence before the middle of the eighth century that the Arabs knew of the Hindu figures.

But a little later this changed. In AD 773 a Hindu brought the astronomical textbook of Brâhmagupta to the court of the Calif at Baghdad. It was translated into Arabic and Arab scholars wrote commentaries on it. One of the scholars whose own studies were stimulated by this book was al-Khowârîzmî, from Khoresm, south of the Aral Sea (see Figure 71). He was the first Arab scholar known to explain the use of the Hindu system of numerals. Al-Khowârîzmî wrote

FIGURE 71 The Aral Sea and Khoresm

an *Algebra* and an *Arithmetic* around AD 820. The *Algebra* has come down to us in two versions, the original Arabic and a twelfth-century Latin translation, *Liber algebrae et almucabala* by Gerard of Cremona, though in the latter considerable portions of the Arabic version are missing. No Arabic version of the *Arithmetic* survives. However, there is what seems to be a fairly faithful translation in the *Algoritmi de numero indorum* and in an extended and more elaborate version of the same book, the *Liber algorismi de practica arithmetice*, both by John of Seville, and also a Latin translation by Robert of Chester from the beginning of the twelfth century.

Al-Khowârîzmî's *Arithmetic* is the first Arabic work in which the decimal place-value system and the computing operations based on it are explained. It is possible to deduce that the original title of the book was something like 'Book about addition and subtraction according to the Hindus' method of computation'. The work begins with the praises of Allah, as was usual at that time, and then continues:

> We have decided to give a picture of the Hindus' method of computation with the aid of nine letters and to show how they produce all their numbers from these for the sake of simplicity and brevity, in order to make it easier for those learning arithmetic; that is, large and small numbers and everything to do with them, multiplication and division, also addition and subtraction.... The Hindus had, then, nine letters, which looke like this: [987654321]. There are variations in the forms of the letters as written by different people, especially, in the case of the 5, 6, 7 and 8, but that does not cause difficulties. They are symbols for denoting a number, and these are the forms where the variations mentioned occur: [5678].

The figures in square brackets are missing in the manuscript, being added later. Unfortunately, it is therefore uncertain which forms al-Khowârîzmî used. At other points in the manuscript there are new figures for 1, 2, 3 and 5 and a circle for zero, but these forms do not correspond to the standard East Arabic type but to those symbols which were current in Western Europe in the twelfth and thirteenth centuries – so here again there is a revision by the translator or copier. Nevertheless the comment that the figures 5, 6, 7 and 8 were written in various ways seems to indicate that al-Khowârîzmî used the East Arabic figures while being aware of the West Arabic figures also. These do differ, in particular in these four cases, as will be shown later.

Al-Khowârîzmî then explains in great detail how the value of a numeral changes when it is put in another place. He was fully aware of the essentials of the place-value system, including the zero. Only the pronunciation of the numerals remains very complicated. As an example he introduces the gigantic numeral 1 180 703 051 492 863 (although the way of writing down the numerals is not given in the manuscript) which he reads in the following way:

> one thousand thousand thousand thousand thousands five times and one hundred thousand thousand thousand thousands four times and eighty thousand thousand thousand thousands four times and then seven hundred thousand thousand thousands three times and three thousand thousand thousands three times and fifty-one thousand thousands twice and four hundred thousands and ninety-two thousands and eight hundred and sixty-three.

In this very laborious notation, which held its own for a long time in Arabic and West European literature, three figures are always brought together to form a group. The groups of three are *counted* from right to left, presumably on practical grounds, since when such large numbers are read the word 'thousand' could easily be omitted once in error.

In al-Khowârîzmî's manuscript there are also descriptions of calculation methods according to the Hindu model with both whole numbers and fractions.

From the twelfth century onwards the *Arithmetic* became known in its Latin translation in the West, and this knowledge of the work spread fast. It made an essential contribution to the slow but irresistible penetration into Western Europe from this time of the Hindu numerals and the corresponding calculation methods. But before going further with this, something more must be said about the further development of the Hindu numerals among the Arabs.

In Arabic texts the numerals were originally expressed in words or in the Greek alphabetical forms. Expression in words was retained for many centuries, even in textbooks in which the 'Hindu arithmetic' was taught. Thus in the Arithmetic of Abû-l-Wafâ (*c*. AD 970) and in the arithmetical treatise of al-Karağî (*c*. 1000), which are both directed at practising mathematicians, all numbers are expressed in words. On the other hand, the Greek letters became rarer and went out of use completely during the twelfth century. In the eighth to ninth centuries a specific Arabic alphabetical style emerged, based on the Greek, and this remained in use until the tenth century. At about the same time a place-value system appeared in which the numbers were represented by the so-called East Arabic numerals (see Figure 72), with a special symbol for the zero. The East Arabic numerals are a modification of the Brâhmî numerals. In the course of the centuries they have undergone only minor changes, mainly in the 5 and the zero. The 5 had a form corresponding to a small Latin *b* standing on its head, later taking the form of a small circle. The zero was first represented as a circle and later as a point. The East Arabic numerals are first to be found on some documents from the year 873 or 874 (= 260 by the Moslem calendar). They were retained in various countries (Egypt, Syria, Turkey, Iran, etc.) and are still in use there today.

The so-called West Arabic numerals (see Figure 72) are contemporary with the East Arabic numerals and are found in the western parts of the Arab-speaking world, likewise stemming from the Hindu figures and being the direct forerunners of our western system of writing numbers. They are still in use today in Morocco and were also current in Spain during the Arab occupation.[32]

The common origin of the West and East Arabic numerals is acknowledged. The 2 and 3 can be traced back in the East Arabic form to a corresponding number of horizontal strokes. The 5, 6, 7 and 8 are strongly differentiated, although a common origin may still be detected (apart from the 8). It is possible that the East

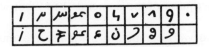

FIGURE 72 Modern East Arabic numerals (top line) and West Arabic numerals from an undated manuscript (bottom line)

Arabic forms came to the Arabs by a circuitous route via Persia. Amendments are easily explained by this hypothesis. Originally the two forms were not very different, they are even found together in some texts – two of the oldest pieces of evidence (from AD 873 and 888) contain numbers in which both forms occur.

The East Arabic numerals are called 'Hindu figures' in Arabic terminology. The West Arabic figures are described as 'gobar'. In Arabic this word means 'sand' or 'dust', and the term bears witness to the fact that these figures were written upon a board strewn with sand such as the Hindus used, and which Maximos Planudes[33] still calls 'Hindu' in the fourteenth century. Stones were not placed on and moved within the marked columns as with the counting board; the number symbols were written in the sand. No zero was necessary for this but dots were placed when needed above the figures. One dot indicated the tens, two dots the hundreds, and so on. For example:

$$\ddot{3}68 = 368, \text{ but } \ddot{3}6 = 360, \ddot{3}6 = 306$$

In this old style the gobar figures do not constitute a complete place-value system, as they do not include a zero. Later the symbol for zero in the form of a small circle was introduced from the East.

There is insufficient evidence to prove without a doubt how and when the Hindu figures came to North Africa and Spain, but it may be assumed that it was a consequence of the trade between the Orient and Moorish Spain. The merchants did not fully understand the novel principles involved in the place-value system with zero but they did see the advantage of writing down a number quickly. Later the pure place-value system from East Arabia gained acceptance. All that was needed, then, was for the zero to join the other nine available figures and for the dots above the figures to disappear.

Figure 73 shows in simplified form the 'family tree' of modern decimal numerals, including the further West European developments of the gobar figures. These are discussed below.

Hindu-Arabic numerals in the West

Spain played a central role in the dissemination of the Islamic intellectual tradition to the Latin-speaking areas of Europe. Spain was, with Sicily, the only European country which had been under the Arabs for centuries and was for that reason especially suited to play a mediating role between Arabs and Christians. Since the West Arabic gobar figures came into use early in Spain it was only a question of time before they would begin to appear in Latin texts. The oldest extant European manuscript with the new figures comes from the Albeldo monastery near the town of Logroño in northern Spain, and dates from AD 976. This so-called *Codex Vigilanus* (see Figure 74), which today is kept in the Escurial, Madrid, has a supplement to the *Origines* of Isidor of Seville (Book 3, Chapter 1) containing a description of the numerals 1 to 9 but without the zero. The preceding text reads:

So with computing symbols. We must realize that the Hindus had the most penetrating intellect and other nations were way behind them in the art of computing, in geometry and in the other free sciences. And this is evident from the nine symbols with which they represented every rank of number at every level. Their forms are as follows: I ⳁ Ʒ ⳋ Ꮍ Ь 7 ⟨ 9

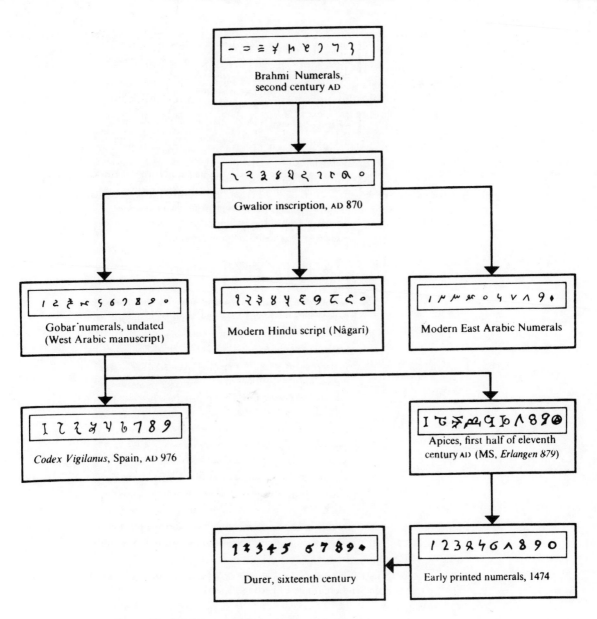

FIGURE 73   Family tree of Hindu-Arabic numerals

A second Spanish manuscript from the year AD 992 (*Codex Emilianus* from San Millan de la Cogolla, near Burgos) gives the same text and the same numerals. The forms show the relationship with the West Arabic numerals and with our present-day forms also.

Thus the Hindu-Arabic numerals came to the West in the first instance through Spain. This was not, however, a continuous process; rather it was only after a number of 'false starts' that the new figures succeeded in establishing themselves in Europe. Three periods may be distinguished:

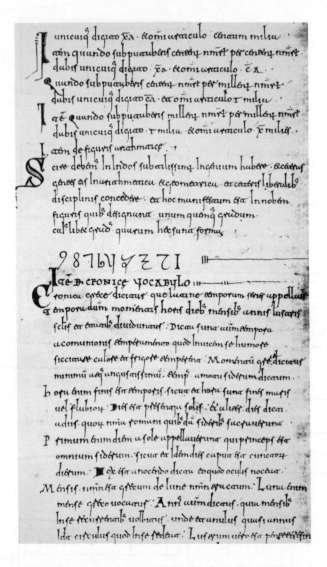

FIGURE 74   Extract from the *Codex Vigilanus* (Real Bibliotheck
de San Lorenzo de El Escorial, Spain)

1. the introduction of the gobar numerals on the abacus by Gerbert and other
   abacists (around 980–1100);
2. the translation of algorithms from the Arabic and the writing of learned
   dissertations on arithmetic (twelfth to thirteenth centuries);
3. the final introduction 'from below' by merchants and practitioners (fifteenth to
   sixteenth centuries).

Gerbert and other abacists

The merit of being the first person to have used the gobar numerals on the western
side of the Pyrenees falls to Gerbert of Aurillac (940–1003), later Pope Sylvester II.

Since he is the most important post-Classical Western mathematician before the twelfth century, we shall briefly discuss his life and his significance for the mathematics of the Middle Ages.

In spite of a lowly background Gerbert obtained a good education in a monastery and rose to high ecclesiastical honours. Shortly after 967 he spent a fairly long time as companion to Count Borel of Barcelona in the Spanish Mark, the border region south of the Pyrenees which was loosely federated with the Frankish empire. There Gerbert undertook scientific studies and learned of the Hindu numerals for the first time. From 972 to 982 he taught the subjects of the *quadrivium* (arithmetic, music, geometry, astronomy) at the cathedral school in Rheims. We are fairly well informed as to the subjects and methods of his teaching from the account by his pupil Richer. Later Gerbert was for a short time Abbot of Ravenna, became archbishop of Rheims in 991, and was elected Pope Sylvester II in 999.

Gerbert's main scientific achievements were in mathematics and astronomy. He wrote a work on geometry (in the style of the Roman land surveyors) and studies of calculations on the abacus, whose popularity he re-established. He did not, however, make use of the old Roman abacus but introduced a modified form of it. Much more essential and more significant than the modified exterior, however, was the fact that Gerbert used small counters of a new kind to represent numbers on the counting board. His biographer Richer writes:

> Gerbert went to no less trouble with his teaching of geometry. As an introduction into the subject he had an abacus made by a shield maker, that is, a tablet which in its dimensions was suited to the purpose. The longer side was divided into twenty-seven columns, and on this he arranged nine symbols for the numbers through which every number could be expressed. He then had one thousand characters similar to these made from horn which were supposed to show alternately on the twenty-seven divisions of the abacus the multiplication or division of various numbers. Thus the division or multiplication of the numbers was carried out so concisely that, in the great majority of examples these processes were understood much more easily than they could be if shown in words.

In contrast to the ancient forms of the counting board the numbers in the columns were not represented by small stones but by special counters inscribed with numerals. Counters such as these were called *apices* – the word *apex* meant literally 'the tip of a cone', and came to mean, among other things, a special style of printing.

Since Gerbert does not illustrate in his writings the nine symbols which he used to represent numbers on the abacus, we are able only to learn indirectly from later sources what they looked like. However, there is an illustration of an abacus of the Gerbert school in a manuscript[34] with the new numerals included on it (see Figure 75). They tally with the figures which Bernelinus, a pupil of Gerbert, was using around 1020, and with the figures in the so-called *Geometry II* by Pseudo-Boethius, a work written in the first half of the eleventh century by an unknown scholar from Lorraine which may be traced back directly to Gerbert's treatise on the abacus, and is otherwise in direct line with the Gerbert tradition. In this Geometry the 'Gerbert figures' are listed twice: firstly the nine ciphers 1–9 in the text, secondly all 10 ciphers (including the zero) on the face of the abacus. The tables in Figure 76 show the numerals in manuscripts for the *Geometry II* dating from the eleventh to the fifteenth centuries.

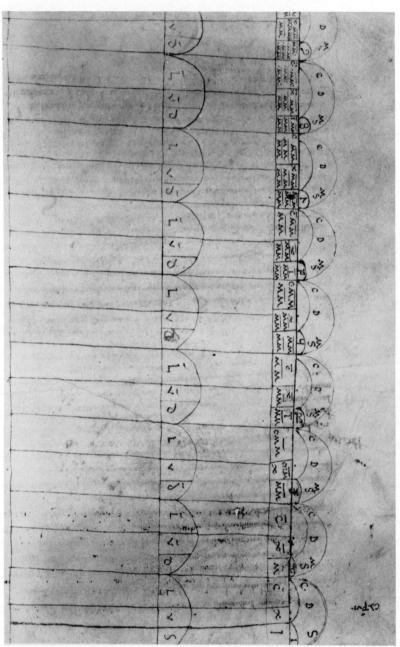

FIGURE 75   An abacus of the Gerbert School (Bibliothèque Nationale, Paris)

Erlangen 379 (eleventh century)
Berlin, lat. oct. 162 (twelfth)
Vat. Barb. lat. 92 (*c.* 1100)
Munich, Clm 23511 (twelfth/thirteenth)
Paris, BN lat. 7377 (*c.* eleventh)
London, BM Harley 3595 (eleventh)
London, BM Lansdowne 842 (fifteenth)
Cesena, Plut. sin XXVI.1 (fifteenth)
London, BM Arundel 339 (thirteenth)
Munich, Clm 13021 (twelfth)
Paris, BN lat. 7185 (twelfth)
Vat. lat. 3123 (twelfth)
Vat. Ottob. lat. 1862 (twelfth)
Chartres 498 (twelfth)

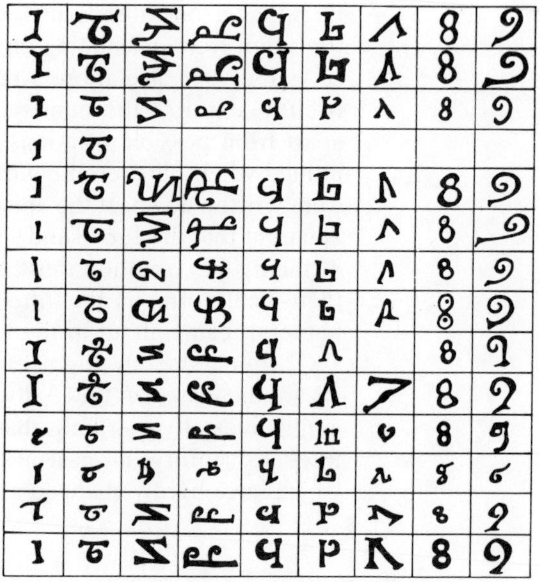

Erlangen 379 (eleventh century)
Berlin, lat. oct. 162 (twelfth)
Vat. Barb. lat. 92 (*c.* 1100)
Munich, Clm 23511 (twelfth/thirteenth)
Paris, BN lat. 7377 (*c.* eleventh)
London, BM Harley 3595 (eleventh)
London, BM Lansdowne 842 (fifteenth)
Cesena, Plut. sin XXVI.1 (fifteenth)
London, BM Arundel 339 (thirteenth)
Munich, Clm 13021 (twelfth)
Paris, BN lat. 7185 (twelfth)
Vat. lat. 3123 (twelfth)
Vat. Ottob. lat. 1862 (twelfth)
Chartres 498 (twelfth)

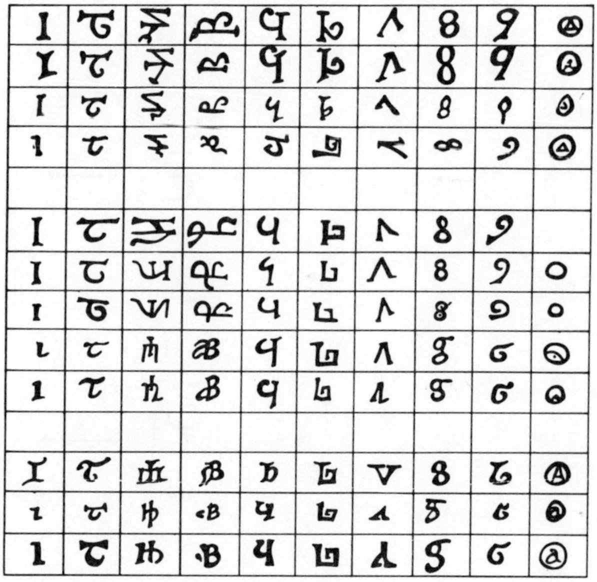

FIGURE 76 Tables of figures in the text of Pseudo-Boethius, *Geometry II*

These two tables show a striking similarity in the forms of the numerals even though the examples range over five centuries. In two manuscripts the 6 has been written as a 7 – presumably mistakes by scribes. Some numerals are rotated through ninety degrees (e.g. the 5, 6, 7 and 8 of the Munich, Clm. 23511); sometimes they are rotated through one-hundred-and-eighty degrees (e.g. the 9 of the last five abacus numerals shown). Thus the form of the numerals was established but not necessarily their orientation, presumably because a counter could be placed on the board with any orientation of the numeral on it.[35]

Eventually, one standard orientation was decided upon, at the latest when the transition from calculating on the abacus to written calculations was made.

As long as calculations were carried out using the Gerbert abacus (up to the twelfth century), the numeral forms of the apices remained virtually unchanged apart from possible rotations through certain specific angles. Later on they were refined when numbered counters were no longer in use but instead the numbers were written down. If the apices are compared with our present-day numerals, it is apparent that the 1, 6, 8 and 9 had almost the modern form even then. With the rest of the figures, rotation must have occurred in the centuries between Gerbert and their final form in the fifteenth century. Probably the development of these numerals came about in the way shown in Figure 77.

Of course the development did not always take place in a direct manner; above all in the case of the 3 the transitional stages are not recognisable without ambiguities. It is striking that the old forms of 4, 5 and 7 are retained into the fifteenth century, at least in Central Europe. Then around 1500 the new forms, which have hardly changed since, became established almost overnight.

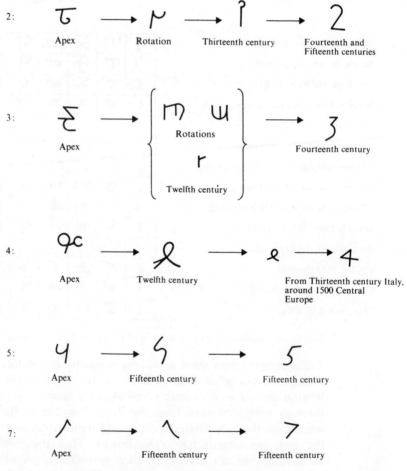

FIGURE 77   Development of the numerals 2, 3, 4, 5 and 7

There were special names for the nine apices which Gerbert used on his abacus, which recurred for the first time in the *Geometry II* of Pseudo-Boethius (shortly after 1000), and which remained unchanged with most of the authors concerned with abacus calculations in the eleventh and twelfth centuries. They were called:

1	2	3	4	5	6	7	8	9
igin	*andras*	*ormis*	*arbas*	*quimas*	*caltis*	*zenis*	*temenias*	*celentis*

There are various hotly disputed hypotheses concerning the origin of these words, tracing them back to Arabic or Greek roots. The names for 4, 8 and 9 are apparently of Arabic origin.

In addition to the nine apices the abacists used an unmarked round counter which was called *sipos* or *rotula*. The word *sipos* is a corruption of the Greek word *psephos* ('counter'). This unmarked counter was used simply as a marker during stages of a calculation. It was placed in a column on the counting board which was temporarily empty, but without any specific thought of the place-value concept. It was of purely technical significance, shifted from column to column as required as an aid to the memory. Radulph of Laon (1131) discusses it as follows:

> The figure, named sipos, does not signify a number but is of service for certain other purposes.... The careful abacist will make for himself among the other symbols on his counters a sipos shaped like a small wheel.

This marker was used in multiplication of large numbers to indicate the figure of the multiplier currently being operated with. A second *sipos* was moved similarly along the multiplicand.

The type of abacus propagated by Gerbert and his school did not become established in Western Europe. The main reason for this was that the substitution of marked apices for unmarked stones did not make calculation clearer or simpler, so that later there was a return to unmarked counters. Gerbert and his school took up the new style from the Arabs without appreciating its advantages, which are only apparent in written calculation. The complete system of writing numerals which the Arabs already possessed was taken over in the West in the early Middle Ages, but blindly, as it were. Its value was not perceived because it was applied to the wrong object, the abacus. The numerous works on the abacus written in the eleventh and twelfth centuries (by, among others, Bernelinus, Gerlandus, Hermann von der Reichenau, Odo, Radulph of Laon and Adelhard of Bath), remained restricted on the whole to the monastery schools. Moreover, written arithmetic developed in opposition to the use of an abacus of this kind. Nevertheless, the abacists did promote knowledge of the Hindu-Arabic system of writing numbers and of the new numerals.

Translation of algorithms and dissertations of arithmetic

From the middle of the twelfth century familiarisation with Latin translations of Arabic books on arithmetic was of decisive importance for the dissemination of the decimal place-value system and of the Hindu-Arabic numerals in Europe. Among these, the *Arithmetic* of al-Khowârîzmî takes pride of place. The Latin treatise of John of Seville on 'practical arithmetic', the Latin translation of Savasorda's *Book on measurements*, and a work entitled *Book introducing algorism into the art of astronomy compiled by Magister A* also played a significant part. Al-Khowârîzmî's name was used in its Latinised form *algorithmus* (or *algorismus*) to denote the 'new' arithmetic, and the word 'algorithm' has been retained to this day with an even wider sense. Around 1200 people started to talk about 'algorists' or 'algorithmists', that is, the exponents of algorithmic arithmetic, as opposed to the 'abacists', that is the exponents of calculation on the abacus. There was a perceptible increase in the number of works on algorism (see Figures 78 to 81); they appeared in various countries, first in Latin but later in the vernacular. Only some of the particularly important treatises can be named here.

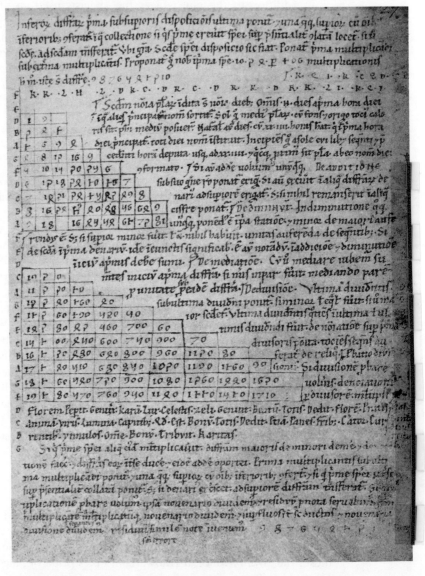

FIGURE 78   Extract from a twelfth-century German manuscript on algorism, showing a multiplication table from 1×1 to 1×9 (Osterreichische Nationalbibliothek, Vienna)

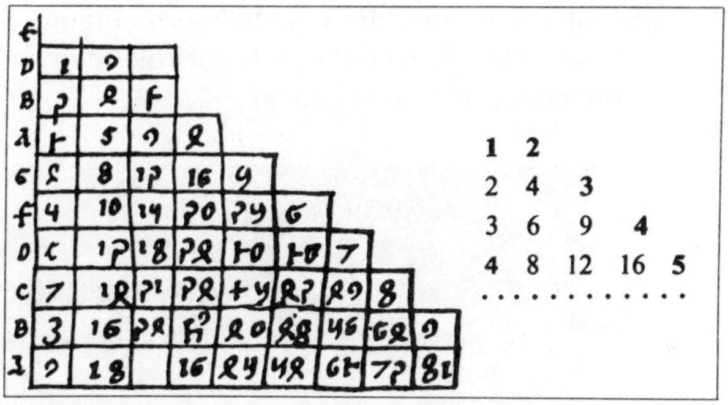

FIGURE 79   The twelfth-century multiplication table explained

The *Liber algorizmi*, which appeared not later than 1200, is one of the oldest writings on algorism. It was found in the Salem monastery on the Bodensee (on the Swiss–German border) and is now preserved in Heidelberg University Library. Other anonymous writings from around 1200 have also been preserved. Algorithmic treatises by Jordanus Nemorarius (thirteenth century), which are records of his lectures, were widely circulated in many manuscripts. The *Common Algorism* (*Algorismus vulgaris*) (1256) of the Englishman Johannes de Sacrobosco (John of Halifax or Hollywood), who had studied in Oxford and went to Paris around 1230 where he taught astronomy and mathematics at the university, enjoyed particular popularity. His book is one of the most successful mathematical works of the Latin Middle Ages; several hundred manuscripts of it are preserved and up until 1582 the work was frequently published, although mathematics had by then advanced considerably.

Around 1290 Petrus de Dacia, a Dane, wrote a well known commentary on Sacrobosco's *Algorism*. The *Song of Algorism* (*Carmen de algorismo*) (c. 1240) by the French mathematician Alexandre de Villedieu, from Normandy, was a comparable success. Calculation with whole numbers is here described in verse in 284 hexameter lines. There are translations of this work into French, English and Icelandic. The earliest 'algorism' in the English language carried the title *Crafte of Nombrynge* (see Figure 80) and appeared around 1300. The works of Sacrobosco

FIGURE 80   First page of the *Crafte of Nombrynge* (British Library)

and of Alexandre de Villedieu (see Figure 81) in particular were widely disseminated and played a significant role in encouraging the teaching of algorithmic
arithmetic in the universities.

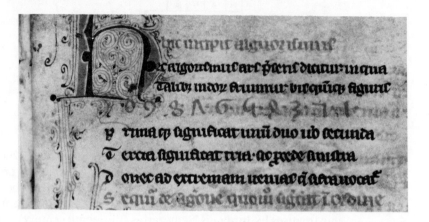

FIGURE 81   Extract from the *Carmen de algorismo* of Alexandre de Villedieu. This
thirteenth-century manuscript presents new computations with numerals in verse from
(Hessische Landes-und-Hochschulebibliothek, Darmstadt)

All these writings on algorism begin with a description of numbers with the aid of
the Hindu-Arabic numerals and then teach the elementary computing operations.
Hindu-Arabic numerals occurred in Europe for the first time shortly before 1000,
but they played only a brief subsidiary role in the West because they were not
suited to calculation on the abacus. The second introduction of the numerals by the
algorists was not crowned with success either. This is certainly to do with the fact
that the algorithmic writings were learned treatises also, compiled by authors with a
scientific interest, and they therefore did not penetrate to the common people
because of their mode of exposition. The written transmission of these works took
place through monks and the content of the writings remained exclusively academic
learning for a long time, so the material was not taken up and used for the everyday
needs of ordinary people. For calculation the people used the abacus as they had
always done (not Gerbert's, but the simpler one with unmarked stones) whereby
calculations could be carried out more slowly but to more obvious effect. At this
point arises a basic obstacle to the introduction of the 'new' system of writing
numbers, namely the zero, which had acquired a symbol although it signified
nothing.

The problem of 'zero' and the final introduction 'from below'

The Sanskrit name for zero is *sunya*; the Arabic name is *as-ṣifr*. Both words have
the same signification: 'the empty'. The Arabic word *ṣifr* was transcribed into
medieval Latin as *cifra* or *zefirum*. However, these two Latin words acquired quite
different meanings. The word *zefirum* (or *cefirum*, as Leonardo of Pisa wrote it in
the thirteenth century) retained its original meaning 'zero'. In Italian it was
changed to *zefiro, zefro* or *zevero*, which was shortened in the Venetian dialect to

zero. On the other hand the word *cifra* acquired a more general meaning: it was used to denote any of the ten signs 0, 1, 2, ..., 9. Hence the French word *chiffre*, and the English word *cipher*. The old English expression 'he is a mere cipher', meaning 'he counts for nothing', still shows that the original meaning of *cipher* was 'zero'. There is an example of its use in the more general sense in the essay *Of Ambition* by Francis Bacon (1561–1626):

He that plots to be the only figure among ciphers, is the decay of the whole age.

A century later in *The Bastard* by Richard Savage (*d.* 1743) the word is used to signify 'nothing' or 'nobody':

Perhaps been poorly rich, and meanly great,
The slave of pomp, a cipher in the state.

Dyche's *New General English Dictionary* of 1760 cites as the only arithmetical meaning:

the mark ... which is expressed by an (0), and which of itself signifies nothing;...

In French, there is the same ambiguity. The original meaning of *chiffre* was 'zero', but the same word was also used to denote the Hindu numerals in general. A French arithmetic textbook for merchants, dated 1485, says:

Et en chiffres ne sont que dix figures, des quelles les neuf ont valeur et la dixième ne vaut rien mais elle fait valloir les autres figures et se nomme zero ou chiffre.

('And of the ciphers there are but ten figures, of which nine are of value and the tenth is worth nothing but gives value to the others and is called zero or cipher'.)
   A similar statement is found in John of Hollywood's *Algorismus* (1256):

Know that corresponding to the 9 limits 9 figures were invented, signifying the 9 digits. They are
   0 9 8 7 6 5 4 3 2 1
The tenth is called *theca* or *circulus* or *cifra* or *figura nihili*, because it stands for 'nothing'. Yet when placed in the proper position, it gives value to the others.

The words *theca* and *circulus* in the extract above refer to the circular form of the sign zero. (*Theca* was the circular brand that was burnt into the forehead of criminals in the Middle Ages.) The word *figures* is still used in English as a synonym of 'numerals'.
   Essentially the confusion over the status of zero is that in one sense the zero was regarded as a 'figure' whilst in another sense it was not. It was also sometimes called *nulla figura* ('no figure'), and yet it was a cipher – it was even *the* cipher par excellence, the one whose original name *cifra* was afterwards given to all the Hindu-Arabic numerals! The zero by itself meant 'nothing', but when placed after another numeral the value of this numeral was enhanced tenfold. This was difficult to understand. As late as the fifteenth century, the zero was described as 'a symbol that merely causes trouble and lack of clarity'.
   A reflection of these medieval doubts and worries is found in *King Lear*, where

the Fool says to the King:

> Now thou art an 0 without a figure. I am better than thou art now. I am a fool, thou
> art nothing.   (Act 1, Scene iv)

The algorists did not succeed in popularising the Hindu-Arabic numerals and the
'new' arithmetic which went with them, so the common people continued to use the
cumbersome Roman numerals even in calculations. As late as 1514, the German
Rechenmeister Ködel published a book on calculation in which he used Roman
numerals only. He called these numerals *gewöhnlich teutsch Zal* ('ordinary German
numbers'). The Hindu-Arabic numerals could only achieve their 'final triumph'
when bankers, traders, merchants and others for whom arithmetical calculations
were part of the daily task became aware of, and adopted, the new system. The
region of highest population density and fullest urban development in medieval
Europe was central and northern Italy, and there the early ascendancy in trade and
commerce was to provide fertile ground for the growth of popular interest in the
decimal place-value system. Italy was also, like Spain, a country which had direct
trading links with the Arabs.

One of the most influential arithmetic textbooks of the period was Leonardo of
Pisa's *Liber abaci* (1202). In this work Leonardo employed systematically and
exclusively the *figurae Indorum* which he had learned as a boy from a Moorish
teacher. He considered all other methods of calculation, including the abacus of
Gerbert, as inferior in comparison with the *modus Indorum*. Leonardo's book
scored a great success. The large merchants realised very soon the great advantages
of the new numbers and they used them in their bookkeeping.

However, in 1299, the city of Florence issued an ordinance which prohibited the
writings of numbers in columns, as well as prohibiting the use of Hindu-Arabic
numerals. The reason for this is supplied by a Venetian treatise on bookkeeping: it
is easy to change a 0 to a 6 or 9. It was less easy to falsify Roman numerals.

The Hindu-Arabic numerals came into general use in Europe via Italy during the
fifteenth century along with such Italian words as *giro, lombard* and *bankrupt*
(*banco rotto*). In 1494 the burgomaster of the German town Frankfurt cautioned
the clerks to use these numerals sparingly. In 1482, a German treatise on
arithmetic by Ulrich Wagner taught 'calculation on lines and with ciphers', i.e. the
use of the abacus and the Hindu numerals. These were the days of passionate argument between the abacists and the algorists. A picture from the *Pearl of Philosophy*
of Gregor Reisch (1504) shows this quarrel in an amusing way. On the right of
Figure 82 is Pythagoras – thought to have been the inventor of the abacus – with
the numbers 1241 and 82 on a counting board. On the left Boethius[36] studies a
calculation in the Hindu-Arabic numerals. Dame Arithmetica is to judge between
them. The sour look on Pythagoras' face suggests that he is the loser!

In the end, the Hindu-Arabic numerals were bound to conquer, though it took a
few more centuries for the conquest to become complete or perhaps one should say
'almost complete' since, even today, it is not difficult to find Roman numerals still
in use (e.g. for chapter numbers in books, copyright years of films, etc.). The
Hindu-Arabic numerals are now, however, in general use in every continent,
though the abacus is still used for calculation in central Asia and in the Far East.

The Hindu-Arabic system was 'bound to conquer' other systems. This is because

FIGURE 82 Woodcut from Reisch, *Margarita philosophica*, 1508, showing Pythagoras (right) using an abacus in the form of a table, competing with Boethius (left) using algorithms and arithmetical signs and numbers to achieve speedy calculations (Ronan Picture Library and the Royal Astronomical Society)

it is a fully cipherised place-value system, but not necessarily because it is based on 10. The Sumerian-Babylonian system was a place-value system based on 60, but it had only two symbols for numbers and thus had to employ the tedious and inconvenient principle of repetition as well as having special problems due to the late invention of a zero. The Greeks had a fully cipherised decimal system, but it was not a place-value system and, for practical purposes of calculation, had certain disadvantages not shared by the Babylonians. The essential ingredients for the 'best' system are that it should be a place-value system and be fully cipherised, that is that the number of distinct numerals (including a zero) should be the same as the base of the system. Only the Hindu system as developed from the sixth century AD met these requirements fully. Whether or not a system based on 10 is the 'best' possible system is still a matter of dispute – a number of mathematicians still argue in favour of a duodecimal (base 12) place-value system requiring 11 distinct numerals together with a zero. The argument is based entirely on the divisibility of the base.

The final two chapters extend the discussion of systems of numerals to fractions and consider methods of calculation with whole numbers and fractions, including calculation using an abacus or counting board and the evolution of the modern computer (see Figure 83).

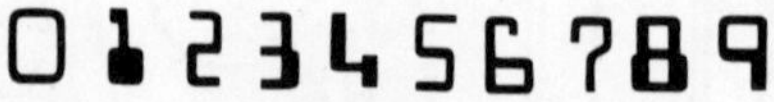

$$0123456789$$

FIGURE 83   Modern numerals as used on bank cheques; a new departure for the benefit of the computer

5 Fractions and Calculation

Natural and unit fractions

In spoken English certain simple fractions like $\frac{1}{2}$ or $\frac{1}{4}$ or $\frac{3}{4}$ have special names. We do not say 'one-fourth' but *one quarter*, and $\frac{1}{2}$ is pronounced as *one half*. The French have a special name *tiers* for $\frac{1}{3}$. Fractions of this kind, which serve the purposes of everyday life, may be called *natural fractions*.

In ancient Egypt the situation was similar. The Egyptians had special words and special symbols for the natural fractions $\frac{1}{2}, \frac{1}{3}, \frac{2}{3}, \frac{1}{4}$ and $\frac{3}{4}$. The symbols for $\frac{1}{2}$ and $\frac{1}{4}$ were

$$\frown = \frac{1}{2} \quad \text{and} \quad \times = \frac{1}{4}$$

An ancient symbol for $\frac{3}{4}$ was replaced later on by the partition

$$\frown \times = \frac{1}{2} + \frac{1}{4} = \frac{3}{4}$$

The fraction $\frac{2}{3}$ had a name of its own: 'the two parts'. It was denoted by a special hieroglyph $\overline{\cap}$. The hieroglyph *r* in the form of a mouth means 'part' and the two strokes mean 'two'. The complement $\frac{1}{3}$, which completed the two parts to a whole unit, was naturally called 'the third part'. Its hieroglyph was $\overline{\cap}$, which means 'the one part', or in later times $\overline{\sqcap\sqcap}$, which may be read as 'the third part'.

In Greek one also speaks of:

'the two parts' $\frac{2}{3}$ 'the third part' $\frac{1}{3}$

'the three parts' $\frac{3}{4}$ 'the fourth part' $\frac{1}{4}$

It presents quite naturally a concrete image: three parts and then a fourth part combine to make the whole. Similarly, this explains the use of the words 'third', 'fourth', 'fifth', etc. In this representation the fifth part is the last part, which combines with the four parts to complete the unit. Etymologically it does not make sense to speak of two fifths, because there is only *one* fifth part, namely the last.

In addition to their notation for natural fractions the Egyptians had a standard notation for unit fractions. To denote a unit fraction $\frac{1}{n}$, they wrote the denominator n below the hieroglyph $r = part$ as follows:

$$\overset{\frown}{\underset{\mathrm{II}\cap}{}} = \frac{1}{12}$$

Following O. Neugebauer, the unit fraction $\frac{1}{n}$ will be denoted by $\bar{n}$, and the

Egyptian natural fraction $\frac{2}{3}$ by $\bar{\bar{3}}$.

As explained in Chapter 4, much of our knowledge of Egyptian mathematics is derived from the Rhind Mathematical Papyrus[1] (see Figure 84). Another highly important source is the Mathematical Leather Roll (see Figure 85), also in the

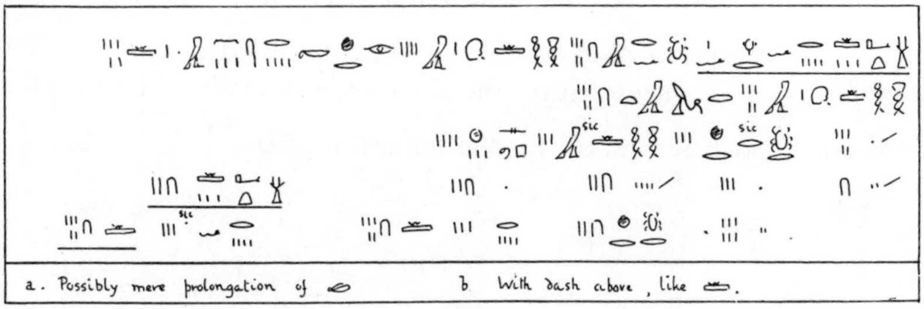

FIGURE 84   Problem 26 of the Rhind Mathematical Papyrus (reproduced from T. E. Peet, *The Rhind Mathematical Papyrus*, University of Liverpool Press, 1923)

British Museum. The Leather Roll is, however, a document of a quite different character. It looks more like a pupil's notebook. It is said to have been found with the Rhind Papyrus near the Ramesseum at Thebes. Its unrolling was a clever application of modern chemistry. At first sight, the contents seemed disappointing: the roll contained only simple identities between fractions such as

$$\frac{1}{5} + \frac{1}{20} = \frac{1}{4}$$

Yet the roll has proved valuable, because it shed some light on the calculations of the Rhind Papyrus.

Both texts, the papyrus and the roll, were written in hieratic script, which is a simplification of hieroglyphic writing. In Peet's publication, the text of the Papyrus is rewritten in hieroglyphs.[2]

The Leather Roll contains identities between the simplest fractions such as

$$\bar{6} + \bar{6} = \bar{3} \text{ and } \bar{6} + \bar{6} + \bar{6} = \bar{2}$$

which are immediately evident. From these, other simple identities like

$$\bar{3} + \bar{6} = \bar{2} \text{ and } \bar{2} + \bar{3} + \bar{6} = \bar{1}$$

can be derived, which are constantly used in the Rhind Papyrus. Dividing the latter

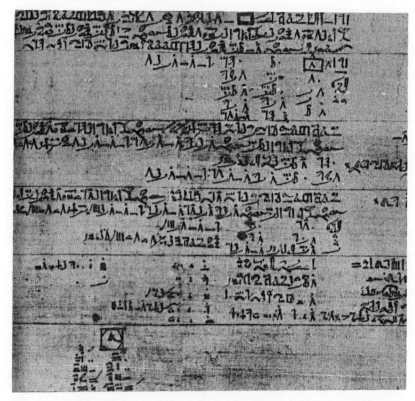

FIGURE 85   Extract from the Mathematical Leather Roll containing simple relations between fractions (by courtesy of the Trustees of the British Museum)

identity by 7, by 9, by 11 and by 15, gives the sequence

$$\overline{14} + \overline{21} + \overline{42} = \overline{7}$$
$$\overline{18} + \overline{27} + \overline{54} = \overline{9}$$
$$\overline{22} + \overline{33} + \overline{66} = \overline{11}$$
$$\overline{30} + \overline{45} + \overline{90} = \overline{15}$$

which is found in the Leather Roll. Similarly, from $\overline{3} + \overline{6} = \overline{2}$ there can be derived relations of the form

$$\overline{3n} + \overline{6n} = \overline{2n}$$

which are also given in the Leather Roll. From these facts one may conclude that *the serial derivation of other identities by division from one fundamental identity was a common procedure in Egyptian arithmetic.*

Egyptian calculation with fractions

In the Rhind Papyrus a divison m/n is usually introduced by the words 'reckon with n so as to obtain m'. Thus, in Problem 24, pupils are asked to 'reckon with 8 so as to obtain 19' (i.e. divide 19 by 8) and in Problem 25 to 'reckon with 3 so as to obtain 16' (i.e. divide

16 by 3). The operations are carried out as follows:

(Problem 24)	19:	8	(Problem 25)	16:	3
	1	8	/	1	3
/	2	16		2	16
	$\bar{2}$	4	/	4	12
/	$\bar{4}$	2		$\bar{3}$	2
/	$\bar{8}$	1	/	$\bar{3}$	1

Quotient: $2 + \bar{4} + \bar{8}$ Quotient: $5 + \bar{3}$

In Problem 24, the number 8 is first doubled to give 16. To get 19, 3 is needed. Half of 8 is now taken to give 4; half of this gives 2, and half again gives 1. Adding the fourth and the eighth parts gives exactly the 3 needed. The final result is obtained by adding the three numbers indicated by strokes, i.e. $2 + \bar{4} + \bar{8}$.

Instead of the sequence $\bar{2}, \bar{4}, \bar{8}, \ldots$ the Egyptians often used the sequence $\bar{\bar{3}}, \bar{3}, \bar{6}, \ldots$. Thus, in Problem 25, after having multiplied the divisor 3 by 2 and by 4, the author of the papyrus takes $\frac{2}{3}$ and next $\frac{1}{3}$ of 3.

It is a peculiarity of the Egyptian calculators not to write down $\bar{3}$ first, but to obtain it from $\bar{\bar{3}}$ by halving. It seems that the pupils were taught always to form $\bar{\bar{3}}$ first.

In Problems 3, 4, 5 and 6 of the Papyrus, we are asked to divide 6, 7, 8 and 9 loaves of bread among 10 men. The answers given are:

$\bar{2} + \overline{10}$ each, $\bar{\bar{3}} + \overline{30}$ each, $\bar{\bar{3}} + \overline{10} + \overline{30}$ each, $\bar{\bar{3}} + 5 + \overline{30}$ each

These answers can be obtained by the Egyptian division method as follows:

	6:	10		7:	10		8:	10		9:	10
	1	10		1	10		1	10		1	10
/	$\bar{2}$	5	/	$\bar{\bar{3}}$	$6 + \bar{\bar{3}}$	/	$\bar{\bar{3}}$	$6 + \bar{\bar{3}}$	/	$\bar{\bar{3}}$	$6 + \bar{\bar{3}}$
/	$\overline{10}$	1		$\overline{10}$	1	/	$\overline{10}$	1		5	2
			/	$\overline{30}$	3	/	$\overline{30}$	3		$\overline{10}$	1
									/	$\overline{30}$	3

In performing multiplications and divisions with unit fractions, the problem often arises to double a unit fraction. If the denominator is even, the solution is trivial: twice $\bar{8}$ is $\bar{4}$. If the denominator is 3, no problem arises either: twice $\bar{3}$ is $\bar{\bar{3}}$. But if the denominator is an odd number such as 5 or 7, the Egyptian scribe faced a serious problem: he was not allowed to write $\bar{5} + \bar{5}$, or $\bar{7} + \bar{7}$. This was probably because repeated doubling would yield such cumbersome expressions as $\bar{7} + \bar{7} + \bar{7} + \bar{7}$ or even worse. In such a situation the scribe used a table which enabled him to write $\bar{5} + \bar{5}$ in a more convenient form as $\bar{3} + \overline{15}$, and $\bar{7} + \bar{7}$ as $\bar{4} + \overline{28}$, and so on. This table is known as the (2:n) table. Using this table, the scribes were able to multiply 7 by 2 or 4 by 8 by repeated doubling as follows

1	$\bar{7}$
2	$\bar{4} + \overline{28}$
4	$\bar{2} + \overline{14}$
8	$1 + \bar{7}$

Right at the beginning of the Rhind Papyrus there is a table of divisions of 2 by odd numbers n, from $n = 3$ up to $n = 101$. The table begins:

What part is 2 of 3? **$\bar{3}$** is 2.

This means: $\dfrac{2}{3}$ of 3 is 2. In this case, a proof is not necessary. The result $\bar{3}$ is written in red ink (shown in bold here). The text goes on:

What part is 2 of 5? **$\bar{\bar{3}}$** is $1 + \bar{3}$, **$\overline{15}$** is $\bar{3}$.

This means: $\dfrac{1}{3}$ of 5 is $1 + \bar{3}$, and $\dfrac{1}{15}$ of 5 is $\bar{3}$. These parts together give 2.

This line in itself would be sufficient if the author merely wanted to give a proof of his result

$$2{:}5 = \bar{3} + \overline{15}$$

However, he also wanted to show how the result was obtained. He therefore added a computation in the standard form of an Egyptian division.

Computation:	1	5
	$\bar{\bar{3}}$	$3 + \bar{3}$
/	$\bar{3}$	$1 + \bar{\bar{3}}$
/	$\overline{15}$	$\bar{3}$

It is seen that the division 2:5 was performed by means of the standard sequence $1, \bar{\bar{3}}, \bar{3}, \ldots$ Line 3 says: $\dfrac{1}{3}$ of 5 is $1 + \bar{3}$. What is missing from the desired result 2 is just $\bar{3}$, which is $\dfrac{1}{15}$ of 5. The quotient $\bar{3} + \overline{15}$ appears in red ink in the top line.

If one divides 2 by 7 in the Egyptian manner, using the sequence $\bar{2}, \bar{4} \ldots$ the result is

$$2{:}7 = \bar{4} + \overline{28}$$

Computation:	1	7
	$\bar{2}$	$3 + \bar{2}$
/	$\bar{4}$	$1 + \bar{2} + \bar{4}$
	$\bar{7}$	1
	$\overline{14}$	$\bar{2}$
/	$\overline{28}$	$\bar{4}$

The columns below collect the results of the divisions $2{:}n$ from $n = 3$ up to $n = 21$. The first column contains those entries in which n is a multiple of 3.

$2{:}3 \ = \ \bar{3}$	$2{:}5 \ = \ \bar{3} + \overline{15}$
$2{:}9 \ = \ \bar{6} + \overline{18}$	$2{:}7 \ = \ \bar{4} + \overline{28}$
$2{:}15 = \overline{10} + \overline{30}$	$2{:}11 = \ \bar{6} + \overline{66}$
$2{:}21 = \overline{14} + \overline{42}$	$2{:}13 = \ \bar{8} + \overline{52} + \overline{104}$
	$2{:}17 = \overline{12} + \overline{51} + \overline{68}$
	$2{:}19 = \overline{12} + \overline{76} + \overline{114}$

How were these results obtained?

Looking at the two columns of the table, one may notice a characteristic difference. All decompositions in the left-hand column are of exactly the same type:

$$2{:}3m = \overline{2m} + \overline{6m}$$

These divisions can be derived from just one fundamental formula

$$\overline{3} + \overline{3} = \overline{2} + \overline{6} \tag{A}$$

by dividing it by 3, by 5, by 7, etc. For example,

if $m = 3$
then $2{:}3m = \overline{2m} + \overline{6m}$
gives $2{:}9 = \overline{6} + \overline{18}$

which is the second row in the left-hand column. Now this formula (A) is applied time and again in many calculations of the Papyrus and the method of serial derivation was a common procedure in Egyptian arithmetic. It is, therefore, a plausible hypothesis that all division results $2{:}3m$ were derived serially from the fundamental formula (A)[3]. In the whole $(2{:}n)$ table divisions $2{:}3m$ are *always* performed by using the formula

$$2{:}3m = \overline{2m} + \overline{6m}$$

To every one of the divisions $(2{:}n)$ the Papyrus gives a 'calculation' of the same kind as that of $(2{:}5)$. Neugebauer has given good reasons to assume that in the cases $2{:}3m$ (left-hand column) these calculations were only verifications added afterwards. The original method was probably serial derivation from one fundamental identity (A).

In the right-hand column, there is a different situation. The six results written in the table are completely different in structure: they cannot be derived from one fundamental identity. The leading (largest) fractions at the right all belong to one of the two sequences

$$\overline{3} \quad \overline{3} \quad \overline{6} \quad \overline{12}\ldots$$

and

$$\overline{2} \quad \overline{4} \quad \overline{6} \quad \overline{8}\ldots$$

This can be explained if one assumes that the method of discovery of this result was just the division method, carried out by using the two standard sequences $\overline{3}, \overline{3}, \ldots$ and $\overline{2}, \overline{4}, \ldots$. In other words, one assumes that the divisions $2{:}n$, as given in the text, are not just verifications afterwards, but were actually used to obtain the results.

A good example is the computation of $2{:}17$. The problem is: to reckon with 17 until we obtain 2, that is to divide 2 by 17. The computation begins:

1	17
$\overline{\overline{3}}$	$11 + \overline{3}$
$\overline{3}$	$5 + \overline{\overline{3}}$
$\overline{6}$	$2 + \overline{2} + \overline{3}$
$/\ \overline{12}$	$1 + \overline{4} + \overline{6}$

Now the text says 'remainder $\overline{3} + \overline{4}$'. In fact, by adding $\overline{3} + \overline{4}$ to $1 + \overline{4} + \overline{6}$, the following is obtained:

$$1 + (\overline{4} + \overline{4}) + (\overline{3} + \overline{6}) = 1 + \overline{2} + \overline{2} = 2$$

One has to find out by what amount one must multiply 17 to get $\overline{3} + \overline{4}$, i.e. to 'reckon with 17 to obtain $\overline{3} + \overline{4}$'. Obviously, this means multiplying 17 by $\overline{51}$ in order to obtain $\overline{3}$, and by $\overline{68}$ in order to obtain $\overline{4}$, so the result of the division is

$$2{:}17 = \overline{12} + \overline{51} + \overline{68}$$

The text gives the computation in a more elaborate form, showing at the same time how the numbers 51 and 68 are obtained as multiples of 17, as follows:

1	$\overline{17}$	
2	$\overline{34}$	
3	$\overline{51}$	$\overline{3}$
.4	$\overline{68}$	$\overline{4}$

Thus, one can see that the author took great care to show how the single steps in the calculation were carried out. He wanted to show by what method the result $\overline{12} + \overline{51} + \overline{68}$ could be *found*, not simply verified.

The Egyptian word for 'computation' or 'working out' is *śśmt*, which may be pronounced as *seshmet*. It is derived from the verb *śśm*, which means 'to guide'. So the purpose of the author was not only to verify the results, but to give his pupils guidance.

If one divides 2 by 23 in the Egyptian manner, using the sequence $\overline{3}, \overline{3}, \overline{6}, \overline{12}$, the result is

$$2{:}23 = \overline{12} + \overline{276}$$

Computation:	1	23
	$\overline{\overline{3}}$	$15 + \overline{3}$
	$\overline{3}$	$7 + \overline{\overline{3}}$
	$\overline{6}$	$3 + \overline{2} + \overline{3}$
/	$\overline{12}$	$1 + \overline{2} + \overline{4} + \overline{6}$

This must be completed to 2, that is $\overline{2} + \overline{4} + \overline{6}$ must be completed to 1. The remainder is $\overline{12}$ because $\overline{4} + \overline{12} = \overline{3}, \overline{3} + \overline{6} = \overline{2}$ and $\overline{2} + \overline{2} = 1$. Now one reckons with 23 to obtain $\overline{12}$, so 23 is multiplied by 12 in an intermediate calculation:

	1	23
/	10	230
/	2	46

This gives the denominator 276, with the final result:

/	$\overline{276}$	$\overline{12}$

In the division 2:23 the problem arises to complete $\overline{2} + \overline{4} + \overline{6}$ to 1. In the same way, in the division 2:29 the problem arises to complete $\overline{6} + \overline{24}$ to 1. As n gets larger, the divisions 2:n give rise to more and more complicated completion problems. How did the Egyptians manage to complete a given sum of fractions to 1?

In such a case *we* would reduce all fractions to a common denominator D, that is, express them as multiples of $\dfrac{1}{D}$. The Egyptians had a similar method. After having chosen a fraction $\dfrac{1}{F}$ they expressed all other fractions as multiples of $\dfrac{1}{F}$. Sometimes F was chosen as the least common multiple of all denominators, but in other cases F was just the largest denominator occurring among the given fractions. Thus it may happen that some of the other fractions are non-integer multiples of $\frac{1}{F}$. (For instance, if the fraction sum to be completed were $\bar{6} + \overline{28}$ and $\frac{1}{F}$ were chosen as $\overline{28}$, then $\bar{6}$ would not be an integer multiple of $\overline{28}$.)

If the given fractions are expressed as multiples of $\frac{1}{F}$ integer or non-integer numbers are obtained, which Neugebauer has called *auxiliary numbers*. In the Papyrus, the auxiliary numbers are written in red ink below the fractions, here in bold. In the following example (Rhind, Problem 7) auxiliary numbers are used to facilitate multiplication. The problem is 'to multiply $\bar{4} + \overline{28}$ by $1 + \bar{2} + \bar{4}$. The 'computation' goes:

1	$\bar{4}$	$\overline{28}$
	7	**1**
$\bar{2}$	$\bar{8}$	$\overline{56}$
	3 + $\bar{2}$	**2**
$\bar{4}$	$\overline{16}$	$\overline{112}$
	1 + $\bar{2}$ + $\bar{4}$	**4**

Total $\bar{2}$

The new unit $\frac{1}{F}$ is $\overline{28}$, the smallest fraction occurring in the first line. The first fraction $\bar{4}$ is 7 times this unit, hence the bold auxiliary number 7 appears below $\bar{4}$. In the third line all numbers are halved, and the auxiliary numbers too are halved. In the fifth and sixth line the numbers are halved once more. Addition of all auxiliary numbers yields 14, hence the required product is 14 times $\overline{28}$, which is $\bar{2}$.

Problems 21–3 of the Rhind Papyrus are completion problems, 21 being a typical example. This says:

It is said to you: What completed $\bar{\bar{3}} + \overline{15}$ into 1?

                      10   **1**   Total 11: Remainder 4.

The given fractions are written as multiples of $\overline{15}$; thus the auxiliary numbers 10 and 1 are obtained. Their sum is 11, hence one has to complete 11 to 15. As the text says, the remainder is 4. Now the text goes on:

Reckon with 15 to find 4.

	1	15
	$\overline{10}$	$1 + \bar{2}$
/	$\bar{5}$	3
	$\overline{15}$	1

Total 4: then $\bar{5} + \overline{15}$ is what must be added to it.
Therefore $\bar{\bar{3}} + \bar{5} + \overline{15} + \overline{15}$ is complete up to 1.
(Check:) **10** **3** **1** **1**

Problem 22 reads 'what completes $\bar{3} + \overline{30}$ into 1?' This problem can be solved by the same method. One takes 30 as a new unit. The auxiliary numbers 20 and 1 add up to 21, and 30 is needed, so the difference is 9. Reckoning with 30 to find 9 gives

$$\begin{array}{cc} 1 & 30 \\ /\ \overline{10} & 3 \\ /\ 5 & 6 \end{array}$$

therefore $\bar{5} + \overline{10}$ is what must be added to it. Check:

$$\bar{3} + \bar{5} + \overline{10} + \overline{30} = 1$$

$$\mathbf{20\quad 6\quad 3\quad 1}$$

In the $(2:n)$ table red numbers appear only in the division 2:35. The text reads:

What part is 2 of 35? $\mathbf{\overline{30}}$ is $1 + \bar{6}$, $\mathbf{\overline{42}}$ is $\bar{3} + \bar{6}$

$$\begin{array}{ccc} \mathbf{6}\quad 7 & & 5 \end{array}$$

Computation:
$$\begin{array}{cc} 1 & 35 \\ \overline{30} & 1 + \bar{6} \\ \overline{42} & \bar{3} + \bar{6} \end{array}$$

The new unit, by which the fractions $\overline{30}$ and $\overline{42}$ were measured, is $\overline{210}$. The fractions $\overline{30}$ and $\overline{42}$ are 7 times and 5 times the new unit, hence their auxiliary numbers are 7 and 5. The first auxiliary number 6 corresponds to the fraction $\overline{35}$, and in fact it appears just below the number 35. In this case the auxiliary numbers 7 and 5 were not written in red, probably because the fractions $\overline{30}$ and $\overline{42}$ were already written in red.

The idea underlying this calculation seems to be as follows. When one asks: What part 2 is of 35 – an obvious answer would be $\overline{35} + \overline{35}$. The Egyptians did not want a result like $\overline{35} + \overline{35}$; they wanted a sum of fractions with *different* denominators. They now chose a new unit $\overline{210}$, and expressed $\overline{35}$ as 6 times this new unit, hence $\overline{35} + \overline{35}$ as 12 times the new unit. This number 12 could be split into 7 and 5. The numbers 7 and 5 have the particular advantage that they are divisors of 35 and hence of $6 \times 35 = 210$ as well, so that 7 new units can be written as a unit fraction $\overline{30}$, and 5 new units as $\overline{42}$. Hence $\overline{35} + \overline{35}$ could be, written as $\overline{30} + \overline{42}$.

The man who invented this decomposition must have been a very clever calculator who saw that 35 is divisible by 5 and by 7, and who had the excellent idea to choose his denominator $210 = 6 \times 35$ in such a way that $6 + 6$ could be split into 5 and 7.

A similar method was applied in the case $n = 91$, but not in other cases. The question 'What part is 2 of 91?' can be solved by analogy with the calculation above. Noting that 91 is divisible by 7 and by 13 leads to

$$2:91 = \overline{70} + \overline{130}$$

The problems 24–38 of the Rhind Papyrus belong to the class of so-called *hau-calculations* or *aha-calculations*. The Egyptian word 'h', which may be pronounced as *hau* or (better) as *aha*, means 'a quantity', 'a collection'. In the aha-calculations it means an unknown quantity. In fact, the aha-calculations are just like the calculations needed to solve a linear equation in one unknown. Problem 26 is a simple example.

A quantity and a fourth part of it are together 15.

The Egyptian solution begins:

Calculate with 4, of this you must take the fourth part, namely 1; together 5.

Then the division 15:5 = 3 is carried out, and finally a multiplication 4 × 3 = 12. The required 'quantity' is therefore 12, its fourth part is 3 together 15.

The method followed here may well be described as the *method of false assumption*: one starts with an arbitrarily chosen number as the required quantity, in this case 4, because this makes the computation of the fourth part easy. Four and a fourth part of four give 5. But the required result is 15; hence the quantity has to be multiplied by 15:5 = 3.

Frequently the quantity is at first taken to be 1, e.g. in Rhind, Problem 37:

I go three times in a bushel; my third part and a third of my third part, and my ninth part are added to me and I come out entirely [i.e. the bushel is entirely filled]. Who says this?

The answer is determined as follows:

$$
\begin{array}{ll}
1 & 1 \\
2 & 2 \\
\bar{3} & \bar{3} \\
\bar{3}\text{ of his }\bar{3} & \bar{9} \\
\text{his }\bar{9} & \bar{9} \\
\text{sum} & 3 + \bar{2} + \overline{18}
\end{array}
$$

(because $\bar{9} + \bar{9} = \bar{6} + \overline{18}$ and $\bar{6} + \bar{3} = \bar{2}$). Then one divides 1 by $3 + \bar{2} + \overline{18}$:

$$
\begin{array}{ll}
1 & 3 + \ \bar{2} + \overline{18} \\
\bar{2} & 1 + \ \bar{2} + \ \bar{4} + \ \overline{36} \\
/\ \bar{4} & \bar{2} + \ \bar{4} + \ \bar{8} + \ \overline{72} \\
\bar{8} & \bar{4} + \ \bar{8} + \overline{16} + \overline{144} \\
\overline{16} & \bar{8} + \overline{16} + \overline{32} + \overline{288} \\
/\ \overline{32} & \overline{16} + \overline{32} + \overline{64} + \overline{576}
\end{array}
$$

Now the auxiliaries enter the field: the sum $\bar{2} + \bar{4} + \bar{8} + \overline{72} + \overline{16} + \overline{32} + \overline{64} + \overline{576}$ must be completed to 1. Taking 576 as the chosen number F, the auxiliary numbers are obtained by multiplying the fractions by 576, and they show that the last five fractions add up to $\bar{8}$ exactly.

$$
\begin{array}{lllllll}
\overline{72} & \overline{16} & \overline{32} & \overline{64} & \overline{576} & \text{sum} & \bar{8} \\
\mathbf{8} & \mathbf{36} & \mathbf{18} & \mathbf{9} & \mathbf{1} & & \mathbf{72}
\end{array}
$$

So the sum of all these fractions is

$$\bar{2} + \bar{4} + \bar{8} + \bar{8} = \bar{2} + \bar{4} + \bar{4} = \bar{2} + \bar{2} = 1$$

and the result of the division is $\bar{4} + \overline{32}$. To check this result, three times $\bar{4} + \overline{32}$, a third, a third of a third, and a ninth of $\bar{4} + \overline{32}$ are determined; then it is shown by means of auxiliaries that their sum is indeed 1.

The group of aha-calculations also includes the first problem of the Berlin papyrus 6619, whose solution requires even the extraction of square roots.[4] The text is as follows:

A square and a second square, whose side is $\frac{3}{4}$ [in the text, $\bar{2} + \bar{4}$] of that of the first square, have together an area of 100. Show me how to calculate this.

The solution starts with a false assumption:

Take a square of side 1, and take $\frac{3}{4}$ of 1 [this is to say $\bar{2} + \bar{4}$] as the side of the other area. Multiply $\bar{2} + \bar{4}$ by itself, this gives $\bar{2} + \overline{16}$. Hence, if the side of the one of the areas is taken to be 1, that of the others is $\bar{2} + \bar{4}$, and the addition of the areas gives $1 + \bar{2} + \overline{16}$. Take the square root of this; it is $1 + \bar{4}$. Take the square root of the given number 100; it is 10. How many times is $1 + \bar{4}$ contained in 10? 8 times.

From here the text becomes undecipherable, but the rest can be easily guessed at: $8 \cdot 1 = 8$ and $8 \cdot (\bar{2} + \bar{4}) = 6$ are the sides of the required squares.

The aha-calculations constitute the climax of Egyptian arithmetic. The Egyptians could not possibly get beyond linear equations and pure quadratics with one unknown with their primitive and laborious computing technique.

Problem 33 of the Rhind Papyrus is a very difficult aha-calculation. The problem reads:

A quantity whose two-thirds, half and seventh are added to it. Now it becomes 37.

To solve this one has to divide 37 by $1 + \bar{\bar{3}} + \bar{2} + \bar{7}$. The computation begins:

1	$1 + \bar{\bar{3}} + \bar{2} + \bar{7}$
2	$4 + \bar{3} + \bar{4} + \overline{28}$ [because $\bar{7} + \bar{7} = \bar{4} + \overline{28}$]
4	$9 + \bar{6} + \overline{14}$ [because $\bar{\bar{3}} = \bar{2} + \bar{6}$]
8	$16 + \bar{\bar{3}} + \bar{7}$
16	$36 + \bar{\bar{3}} + \bar{4}$ $+ \overline{28}$
28	$10 + \bar{2}$ **$1 + \bar{2}$**

The sum $36 + \bar{\bar{3}} + \overline{28}$ is already quite near to 37. How much is missing? In order to complete $\bar{\bar{3}} + \bar{4} + \overline{28}$ to 1, auxiliary numbers are introduced. In this case the new unit, by which all other fractions are measured, is not the smallest fraction $\overline{28}$, but $\overline{42}$. The number 42 is chosen for the continuation of the calculation because it is important that the sum of the auxiliary numbers belonging to the divisor $1 + \bar{\bar{3}} + \bar{2} + \bar{7}$ be an integer. The new denominator 42 is simply the least common multiple of the denominators 3, 2 and 7.

The auxiliary numbers 28, $10\frac{1}{2}$ and $1\frac{1}{2}$ were obtained by multiplying 42 by the fractions $\bar{3}$, $\bar{4}$ and $\overline{28}$, as follows:

$$
\begin{array}{cc}
1 & 42 \\
/\ \bar{3} & 28 \\
\bar{2} & 21 \\
/\ \bar{4} & 10 + \bar{2} \\
/\ \overline{28} & 1 + \bar{2}
\end{array}
$$

The text goes on:

 sum 40, remainder 2.

In fact, the sum of the auxiliary numbers $28 + (10 + \bar{2}) + (1 + \bar{2})$ is 40, and 42 new units are needed, so 2 new units are missing. Next the divisor $1 + \bar{3} + \bar{2} + \bar{7}$ is also multiplied by 42, which gives 97. Finally, the Egyptian scribe divides 2 by 97. Using the $(2{:}n)$ table, he writes the result as $\overline{56} + \overline{679} + \overline{776}$ and he concludes that the required 'quantity' is

$$16 + \overline{56} + \overline{679} + \overline{776}$$

The line of thought of the Egyptian calculator may be explained as follows. Recall that in order to complete $\bar{3} + \bar{4} + \overline{28}$ to 1 auxiliary numbers were introduced, using $\overline{42}$ as a new unit. The sum $\bar{3} + \bar{4} + \overline{28}$ gives 40 new units, so 2 new units are missing. Hence the remainder R, which completes $\bar{3} + \bar{4} + \overline{28}$ to 1, is equal to 2 new units, i.e. twice $\overline{42}$, or $\overline{21}$. The next step would be to divide $\overline{21}$ by $1 + \bar{3} + \bar{2} + \bar{7}$ or, in Egyptian terminology to reckon with $1 + \bar{3} + \bar{2} + \bar{7}$ to obtain $\overline{21}$. This division is performed by means of auxiliary numbers, the new unit being again $\overline{42}$. Instead of dividing $\overline{21}$ by $1 + \bar{3} + \bar{2} + \bar{7}$ one divides the auxiliary number 2 by the auxiliary number

$$(1 + \bar{3} + \bar{2} + \bar{7}) \times 42 = 97$$

The result of this division 2:97 is, according to the $(2{:}n)$ table, $\overline{56} + \overline{679} + \overline{776}$. Adding this to 16, the required 'quantity'

$$16 + \overline{56} + \overline{679} + \overline{776}$$

is obtained. A very skilful calculation!

 The following two aha-calculations comprise Problems 24 and 34 of the Rhind Papyrus.

1. 'A quantity and a seventh part of it are together 19' (assuming 7 is the answer).
2. 'A quantity, its half and a fourth part are together 10'. To check the answer using auxiliary numbers fractions other than $\bar{2}$ and $\bar{8}$ will be referred to $F = 56$.

The solutions in the Egyptian manner are as follows:

1. Assuming 7 as the answer gives $7 + 1 = 8$. One must therefore divide 19 by 8.

$$
\begin{array}{rl}
1 & 8 \\
/\ 2 & 16 \\
\bar{2} & 4 \\
/\ \bar{4} & 2 \\
/\ \bar{8} & 1 \\
\end{array}
$$

Sum $2 + \bar{4} + \bar{8}$

One multiplies $2 + \bar{4} + \bar{8}$ by 7:

$$
\begin{array}{rl}
/\ 1 & 2 + \bar{4} + \bar{8} \\
/\ 2 & 4 + \bar{2} + \bar{4} \\
/\ 4 & 9 + \bar{2} \\
\end{array}
$$

The solution is thus $16 + \bar{2} + \bar{8}$.

2. Divide 10 by $1 + \bar{2} + \bar{4}$

$$
\begin{array}{rll}
/\quad 1 & 1 + \bar{2} + \bar{4} & \\
2 & 3 + \bar{2} & \\
/\quad 4 & 7 & \\
/\quad \bar{7} & \bar{4} & (\text{since } 4 \times (1 + \bar{2} + \bar{4}) = 7) \\
\bar{4} + \overline{28} & \bar{2} & (2{:}7 = \bar{4} + \overline{28}) \\
/\quad \bar{2} + \overline{14} & 1 & \\
\end{array}
$$

Sum $5 + \bar{2} + \bar{7} + \overline{14}$

The answer is checked by finding $1 + \bar{2} + \bar{4}$ times $5 + \bar{2} + \bar{7} + \overline{14}$.

$$
\begin{array}{rl}
1 & 5 + \bar{2} + \bar{7} + \overline{14} \\
/\ \bar{2} & 2 + \bar{2} + \bar{4} + \overline{14} + \overline{28} \\
/\ \bar{4} & 1 + \bar{4} + \bar{8} + \overline{28} + \overline{56} \\
\end{array}
$$

Sum $9 + \bar{2} + \bar{8} + \bar{7} + \overline{14} + \overline{14} + \overline{28} + \overline{28} + \overline{56}$

$$
\begin{array}{ccccc}
8 & 4 & 4 & 2 & 2 & 1 \\
\end{array}
$$

$$
\begin{array}{cc}
\bar{4} + \bar{8} \\
14 \quad 7 \\
\end{array}
$$

Since both sets of auxiliaries add up to 21, the sum is

$$
9 + \bar{2} + \bar{8} + \bar{4} + \bar{8} = 10
$$

Babylonian sexagesimal fractions and algebra

As the Babylonians had a sexagesimal place-value system for integers and fractions, all numbers were written as sums of powers of 60, multiplied by integers from 0 to 59. For instance, if a Sumerian scribe wanted to write the number $316\frac{2}{3}$, he wrote it as

$$\text{𒐊 𒐏 𒐊 𒐏 𒐊}$$

that is as 5, 16, 40 which means

$$5 \times 60 + 16 + \frac{40}{60}$$

The modern 'decimal point' was missing, so the notation did not indicate by what powers of 60 the figures had to be multiplied. Thus, the same notation 5, 16, 40 might also denote the number

$$5 \times 60^2 + 16 \times 60 + 40$$

or the number

$$5 + \frac{16}{60} + \frac{40}{60^2}$$

and so on.

The following transcription of Babylonian calculations indicates the point of separation between integers and fractions by a semicolon. Thus,

$$316\,\frac{2}{3} \text{ is written as}$$

$$5, 16; 40$$

If the place value of the figures is not known, the semicolon will be omitted and just commas inserted.

The sexagesimal place-value notation was being used by the Sumerians as early as 2000 BC. On Sumerian tablets from the time of the king Shulgi there are tables of reciprocals ($1/x$) and multiplication tables. The latter are either single tables, containing the multiples of a single 'head number', or combined tables consisting of several single tables. Figure 86 shows an example of a single multiplication table. Obviously, *a-rá* means 'times'. Another example – for multiples of 16, 40 – would be:

$$(16, 40)\ \textit{a-rá} \quad 1 \quad 16, 40$$

$$\textit{a-rá} \quad 2 \quad 33, 20$$

$$\textit{a-rá} \quad 3 \quad 50$$

$$\cdots$$

$$a\text{-}r\acute{a} \quad 19 \quad 5, 16, 40$$

$$a\text{-}r\acute{a} \quad 20 \quad 5, 33, 20$$

$$a\text{-}r\acute{a} \quad 30 \quad 8, 20$$

$$a\text{-}r\acute{a} \quad 40 \quad 11, 6, 40$$

$$a\text{-}r\acute{a} \quad 50 \quad 13, 53, 20$$

	7 *a-rá* 1 7
	a-rá 2 14
	a-rá 3 21
	. . .
	a-rá 19 2, 13
	a-rá 20 2, 20
	a-rá 30 3, 30
	a-rá 40 4, 40
	a-rá 50 5, 50

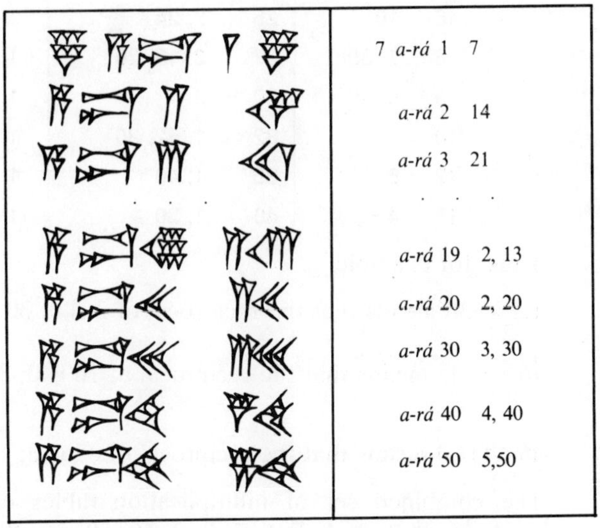

FIGURE 86   Sumerian table of multiples of 7

Frequently several single multiplication tables were combined with a table of reciprocals and a table of squares to form a large combination table. Before one can understand the arrangement of such a combination the tables of reciprocals must first be addressed.

There are large tables of reciprocals dating from the times of the Seleucids (311–1 BC). These were probably designed for astronomers. (They are reproduced in Neugebauer's *Mathematische Keilschrifttexte*.) One of them begins as follows:

1:1	= 1
1:1, 0, 16, 53, 53, 20	= 59, 43, 10, 50, 52, 48
1:1, 0, 40, 53, 20	= 59, 19, 34, 13, 7, 30
1:1, 0, 45	= 59, 15, 33, 20
1:1, 1, 2, 6, 33, 45	= 58, 58, 56, 38, 24.

In this way it continues through seven pages of *Mathematische Keilschrifttexte* up to

$$1:3 = 20$$

The older tables are not so extensive. They usually contain the reciprocals of those integers between 0 and 81, which contain only factors 2, 3 and 5 and which have therefore reciprocals that can be expressed as finite sexagesimal fractions. Part of such a 'normal' table of reciprocals is given below:

1:	2 = 30	16	3, 45	45	1, 20
	3 20	18	3, 20	48	1, 15
	4 15	20	3	50	1, 12
	5 12	24	2, 30	54	1, 6, 40
	6 10	25	2, 24	1	1
	8 7, 30	27	2, 13, 20	1, 4	56, 15
	9 6, 40	30	2	1, 12	50
	10 6	32	1, 52, 30	1, 15	48
	12 5	36	1, 40	1, 20	45
	15 4	40	1, 30	1, 21	44, 26, 40

Thus, for example:

1:2 = 30 means that the reciprocal of 2 is 0; 30 (i.e. 30/60).

16   3, 45 means that the reciprocal of 16 is 0; 3, 45 $\left(\text{i.e. } \dfrac{3}{60} + \dfrac{45}{60^2}\right)$.

45   1, 20 means that the reciprocal of 45 is 0; 1, 20 $\left(\text{i.e. } \dfrac{1}{60} + \dfrac{20}{60^2}\right)$.

The combined set of multiplication tables contains not only tables for the multiples of 2, 3, 4, 5, 6, 7, 8, 9, 10, 20, 30, 40, 50, which are to be expected in any ordinary multiplication table, but also tables for the multiples of several other numbers of two and three digits, by all of the following numbers:

50	24	12	6, 40	2, 30
48	22, 30	10	6	2, 24
45	20	9	5	2, 15
44, 26, 40	18	8, 20	4, 30	2
40	16, 40	8	4	1, 40
36	16	7, 30	3, 45	1, 30
30	15	7, 12	3, 20	1, 20
25	12, 30	7	3	1, 15

What determined the choice of these numbers? Most of them also occur in the normal table of reciprocals and the others, except the number 7, are reciprocals of simple numbers. This also accounts for the arrangement in order of decreasing magnitude; it is indeed the order of the reciprocals in the table of reciprocals.

Thus it can be seen that the multiplication table served not only for ordinary multiplications $a \cdot b$, but especially for multiplications of the form $a \cdot b^{-1}$, i.e for divisions a/b.

The matter can also be stated as follows: the combined tables were used to multiply numbers, but also to represent common fractions as sexagesimal fractions.

For example, to write $\frac{3}{8}$ in sexagesimal form, one first locates $1:8 = 0; 7, 30$ in the table of reciprocals and then the product of this result by 3 in the multiplication table: $0; 22, 30$.

The mathematical texts fully confirm this interpretation. Whenever a division $a:b$ is to be carried out in these texts, one is told (not in terms of a general formula, but for definitely specified numbers): calculate the reciprocal b^{-1} and multiply it by a.

Thus the Sumerian-Babylonian calculation tables were arranged in a very useful manner. A systematic use of the advantages of the positional notation avoids all messing about with fractions; the four rational operations can be carried out rapidly without further thought.

Whenever a division did not go without remainder, approximations were used. An ancient Babylonian text[5] gives approximate values for the reciprocals of all numbers from 40 or 50 to 80 in the following manner:

$\cdots$

```
1:59    = 1, 1, 1
1: 1    = 1
1: 1, 1 = 59, 0, 59
1: 1, 2 = 58, 3, 52
```

$\cdots$

In transcribed Babylonian notation to two fractional places the reciprocals of 14 and 1; 7 are expressed as:

$$0; \quad 4, 17$$

$$0; 53, 44$$

These results may be obtained by long division, converting carrying figures by multiplication by 60. For example:

$$\frac{1}{14} = 0; \ldots$$

To obtain the first sexagesimal fractional place, multiply 1 by 60. Dividing 60 by 14 gives

$$\frac{1}{14} = 0; 4, \ldots$$

with remainder 4. Then multiplying 4 by 60 and dividing by 14 gives

$$\frac{1}{14} = 0; 4, 17, \ldots$$

Very often, a combined multiplication table was followed by a table of squares.

Such a table usually contains the squares of all integers from 1 to 59, as follows:

1   *a-rá* 1   1
2   *a-rá* 2   4
3   *a-rá* 3   9
...
59   *a-rá* 59   58, 1

The last line in the table can be easily checked:

$$59 \times 59 = 3481 = 58 \times 60 + 1$$

(or, more elegantly,

$$59 \times 59 = 60 \times 59 - 59 = 59, 0 - 59 = 58, 1)$$

From such a table of squares, a table of square roots can be immediately derived. The Babylonians had such tables. A standard table of square roots begins:

1-*e*   1   *ib-si* [i.e. of 1 is 1 the root]
4-*e*   2   *ib-si*
9-*e*   3   *ib-si.*

In Babylonian mathematics these tables were used in the solution of quadratic equations. In the same way, tables for cube roots,

1-*e*   1   *ba-si*
8-*e*   2   *ba-si*
27-*e*   3   *ba-si,* etc.

were used for the solution of pure cubic equations

$$x^3 = a$$

The word *ba-si* however does not only mean 'cube-root', but more generally 'root of an equation'. Indeed there are also tables for the equation

$$x^2(x + 1) = a$$

in which this word occurs as well:

2-*e*   1   *ba-si*
12-*e*   2   *ba-si*
36-*e*   3   *ba-si,* etc.

The first column gives the value of $x^2(x + 1)$, and the second column gives the corresponding values of the root x.

Moreover, the same word *ba-si* is found in the text VAT 8521. The museum signature VAT means 'Vorder-Asiatische Texte'. These texts are in the Berlin Museum, now in East Berlin. In the text, the word *ba-si* is followed by 1-*lal*, which means 'less 1'. The expression '*ba-si* 1 *lal*' means: root of the equation

$$x^2(x - 1) = a$$

The numbers $n^2(n + 1)$ and $n^2(n - 1)$, which occur in these texts, also play a part in late Greek texts, such as the *Arithmetical Introduction* of Nicomachus of Gerasa. Nicomachus has a special name for such numbers: he calls them *Arithmoi para-mekepipedoi*, and he gives them an interpretation as volumes of parallelopipeds in which two edges are equal to the integer *n*, while the third edge is either $n + 1$ or $n - 1$. Nicomachus was a Neo-Pythagorean living in the first century AD. In general, his teachings are based upon much more ancient Pythagorean doctrines, which show many points of contact with Babylonian arithmetic and algebra.[6]

Of all these roots, the square roots were the most important, for they were used in solving quadratic equations. Of course, one finds $\sqrt{b}$ in a table of square roots only if *b* happens to be a perfect square. What did the Babylonians do if they had to extract a root from a number *b* which is not a perfect square? They did just what today's engineers do: they used approximations. Thus, a cuneiform text from the Yale Babylonian Collection gives 1; 25, 51, 20 as an approximation for $\sqrt{2}$ (see Figure 87).

FIGURE 87   Cuneiform text (Babylonian Collection, Sterling Memorial Library, Yale University)

The significance of the number 30 together with the numbers 1, 24, 51, 10 and 42, 25, 35 which appear on the tablet can be explained as follows: 30 is the length of the side of the square and 42;25,35 is the length of its diagonal. The ratio between these two lengths is

$$\frac{42; 25, 30}{30} = 1; 24, 51, 10$$

How did the Babylonians obtain such approximations? There is a method for finding approximations to square roots, which is known from Babylonian as well as from Greek texts.

Suppose one wants to find an approximation for $\sqrt{b}$, starting with a rough approximation a, so that a^2 does not differ much from b. After writing

$$b = a^2 + d$$

(the difference d may be positive or negative), one then tries to find a better approximation $a + y$. If the first approximation a is not too bad, y may be expected to be small compared with a. For y there is the condition

$$(a + y)^2 = a^2 + d$$

or

$$2ay + y^2 = d$$

Now y^2 is small compared with $2ay$. If the term y^2 is neglected, the equation

$$2ay = d$$

is obtained which may be solved for y to give

$$y = \frac{d}{2a}$$

Hence, $a + y = a + \dfrac{d}{2a}$ is a better approximation. If necessary, the process can be repeated to obtain better and better approximations.

In transcribed Babylonian notation the second and third approximations for $\sqrt{2}$, starting with $a = 1\frac{1}{2}$ as a first approximation, are as follows.

Since $a^2 = 2\frac{1}{4}$ and $b = 2$, we have $d = -\frac{1}{4}$ and hence

$$y = \frac{d}{2a} = \frac{-\frac{1}{4}}{3} = -\frac{1}{12}$$

The second approximation is therefore

$$a + y = 1\frac{1}{2} - \frac{1}{12} = 1\frac{5}{12}$$

$$= 1 \; ; 25 \text{ in sexagesimal notation.}$$

Calling this number a_1, gives

$$a_1^2 = \frac{289}{144} = 2\frac{1}{144} \text{ and } b = 2$$

hence

$$d_1 = b - a_1^2 = -\frac{1}{144}$$

$$y_1 = \frac{d_1}{2a_1} = -\frac{1}{144} \cdot \frac{1}{2} \cdot \frac{12}{17} = -\frac{1}{24 \times 17}$$

To convert this fraction into the sexagesimal system, multiply it by 60^2, thus obtaining

$$60^2 y_1 = -\frac{150}{17} = -8\frac{14}{17}$$

To convert the fraction $\frac{14}{17}$ into a sexagesimal fraction, again multiply it by 60:

$$60 \times \frac{14}{17} = \frac{840}{17} = 49\frac{7}{17}$$

This result rounded up to 50 gives

$$60^2 y_1 = -8; 50$$

hence

$$y_1 = -0; 0, 8, 50$$

and

$$a_1 + y_1 = 1;25 - 0; 0, 8, 50$$
$$= 1; 24, 51, 10$$

(Note that this is just the Babylonian approximation for $\sqrt{2}$.)

The following discussion of Babylonian algebra will often refer to the standard edition of O. Neugebauer: *Mathematische Keilschrifttexte* (Julius Springer, Berlin 1935), in which all mathematical cuneiform texts known in 1935 are reproduced, translated and commented upon. This standard edition will be quoted as *MKT* I, II or III.

Already in the Old Babylonian period, that is in the time of the dynasty of Hammurabi (1830–1530 BC), the Babylonians had a fully developed algebra. Babylonian algebra was mainly concerned with solving equations and systems of equations. Methods for solving equations were first developed in the Old Babylonian period, and the same methods were still in use 1500 years later in the Hellenistic period under the Seleucid rulers.

Equations and methods of solution were always formulated in words. An unknown quantity x was often called *side* (of a square or rectangle), and its square was called *square*. When two unknowns occurred, they were usually called *length* and *breadth*, and their product *area*. Three unknowns were sometimes called *length*, *breadth* and *height*, and their product *volume*.

Despite this geometric terminology, the Babylonians did not hesitate to subtract a side from an area. Thus, the quadratic equation.

$$x^2 - x = 14,30 \tag{1}$$

was formulated as follows in the text BM (= British Museum) 13901 (see *MKT* III, p. 6, problem 2):

I have subtracted the side of the square from the area, and the result is 14, 30.

The rules for solving equations were always explained as operations on definite numerical coefficients. For example the equation (1) was solved as follows (the words in square brackets are added by way of an explanation):

> Take 1, the coefficient [of the linear term]. Take one-half of 1 [result 0; 30]. Multiply 0; 30 by 0; 30 [result 0; 15]. Add 0; 15 to 14, 30. Result 14, 30; 15. The square root of this is 29; 30. Take 0; 30, which you have multiplied by itself, and add to it 29; 30. Result 30: this is the [required side of the] square.

If one compares the Babylonian method for solving the equation

$$x^2 - x = 14, 30$$

with the modern method of solving an equation

$$x^2 - ax = b$$

by means of the formula

$$x = \frac{a}{2} \pm \sqrt{\left(\frac{a}{2}\right)^2 + b}$$

then in what respect do the two methods agree, and what are the points of difference? In fact, the calculations are exactly the same in both cases, but we use letters a and b for the coefficients, whereas the Babylonians presented single numerical examples. Also we admit two roots, one positive and one negative, whereas the Babylonians admitted only positive roots. The following types of equations occur in cuneiform texts.

1. Linear equations in one unknown

The solution of the equation

$$ax = b$$

was, of course

$$x = a^{-1}b$$

The first factor a^{-1} was taken from a table of reciprocals, and the product $a^{-1}b$ was determined from a multiplication table. If necessary, b was written as a sum of simpler terms. Thus, the number 1,21 would be decomposed into 1, 0 and 20 and 1.

When a^{-1} was not a finite sexagesimal fraction, the Babylonians either used an approximation, or they presented the result x without further comment. Thus the solution of $1; 10x = 5, 50$ was given as $x = 5,0$.

2. Systems of linear equations in two or three unknowns

The usual method of solution is: solve one of the equations for one of the unknowns and substitute into the other equations, etc. However, in some cases a different method was used. Suppose one has to solve an equation of the form $x + y = a$ and another (linear or non-linear) equation for x and y. If x and y were equal, each of them would be $\frac{1}{2}a$. Now suppose x exceeds $\frac{1}{2}a$ by an unknown amount s:

$$x = \frac{1}{2}a + s$$

Then one must take for y

$$y = \frac{1}{2}a - s$$

Substituting these two expressions into the other equation gives an equation for s, which can be solved by Babylonian methods if it is linear or quadratic.

Call this method the *plus and minus* method. The late-Greek writer Diphantus often used this method if a sum of two unknown numbers $x + y$ was given. An example would be to solve the simultaneous equations

$$\left.\begin{aligned} \frac{2}{3}x - \frac{1}{2}y &= 8, 20 \\ x + y &= 30,0 \end{aligned}\right\}$$

by the 'plus and minus' method.

Substituting $x = 15, 0 + s$ and $y = 15, 0 - s$ into the first equation, gives

$$\frac{7}{6}s = 5,50 \tag{1}$$

or, if the coefficient $\frac{7}{6}$ is expressed as a sexagesimal fraction

$$1; 10s = 5, 50 \tag{2}$$

The solution is obvious: $s = 5, 0$. Therefore

$$x = 15, 0 + 5, 0 = 20, 0 \tag{3}$$
$$y = 15, 0 - 5, 0 = 10, 0 \tag{4}$$

These equations are taken from the text VAT 8389. The auxiliary equation (2) is expressed in words: 'What must I multiply by 1; 10 to obtain 5, 50?' The answer is: 'Take 5, 0', since 5, 0 × 1; 10 = 5, 50. Next x and y are calculated by (3) and (4). The text clearly follows the 'plus and minus' method.[7]

3. A linear and a quadratic equation

The following standard types occur frequently:

$$x + y = a, \quad xy = b \tag{1}$$

$$x - y = a, \quad xy = b \tag{2}$$

$$x + y = a, \quad x^2 + y^2 = b \tag{3}$$

$$x - y = a, \quad x^2 + y^2 = b \tag{4}$$

The cases (1) and (3), in which the sum $x + y$ is given, were solved by the 'plus and minus' method. Substituting $x = \frac{1}{2}a + s$ and $y = \frac{1}{2}a - s$ into the second equation, results in both cases in an equation of the form

$$cs^2 = d \qquad (5)$$

from which s can be obtained as a square root.

In the case (2) and (4), were $x - y = a$ is given, a similar method was applied. Putting

$$x = s + \frac{1}{2}a$$

$$y = s - \frac{1}{2}a$$

results in both cases in an auxiliary equation for s of the same form (5).

The general solutions of problems (1) and (2) above can now be worked out.

The solution of (1) is:

$$x = \frac{1}{2}a + s$$

$$y = \frac{1}{2}a - s$$

with

$$s = \sqrt{\left(\frac{1}{2}a\right)^2 - b}$$

The solution of (2) is:

$$x = s + \frac{1}{2}a$$

$$y = s - \frac{1}{2}a$$

with

$$s = \sqrt{\left(\frac{1}{2}a\right)^2 - b}$$

Very often other equations were reduced to the standard types (1) to (4). Thus, to solve the simultaneous equations (*MKT* I, p. 113)

$$\left. \begin{array}{c} xy + (x - y) = 3, 3 \\ x + y = 27 \end{array} \right\}$$

the Babylonians first added the two equations to give

$$xy + 2x = 3, 30$$

They next reduced the pair to the standard form (1) by introducing $y' = y + 2$ as a new unknown to give

$$x + y' = 29 \atop xy' \ \ = 3, 30 \Big\}$$

The solution of this pair of equations,

$$x = 15, \ y' = 14$$

was calculated by the standard method. Finally y, the 'true width' of the rectangle, was found as

$$y = y' - 2 = 12$$

4. Single quadratic equations

Two types occur frequently

$$x^2 - ax = b \tag{1}$$

$$x^2 + ax = b \tag{2}$$

In both cases, the Babylonian method is equivalent to the formula in the first case

$$x = \frac{1}{2}a + \sqrt{\left(\frac{1}{2}\right)^2 + b}$$

and in the second case

$$x = \sqrt{\left(\frac{1}{2}a\right)^2 + b} - \frac{1}{2}a$$

5. Simultaneous quadratic equations

The following example[8] is far from easy:

$$0; 20(x + y) - 0; 1(x - y)^2 = 15 \atop xy = 10 \Bigg\}$$

We do not know how this problem (YBC 4697, Problem D2) was solved by the Babylonians. One possible method would be to put

$$x = u + v$$

and

$$y = u - v$$

thus obtaining

$$0; 40u - 0; 4v^2 = 15 \tag{1}$$

$$u^2 - v^2 \ = 10 \tag{2}$$

Eliminating v^2, gives a quadratic equation for u which can be solved by the standard method.

Multiplying (1) by 15 and subtracting it from (2) gives

$$u^2 - 10u = 6, 15$$

or

$$(u - 5)^2 = 6, 40$$

hence

$$u - 5 = 20$$

or

$$u = 25$$

Substituting $u^2 = 10, 25$ into (2), gives

$$v^2 = 25$$

hence

$$v = 5$$

and finally

$$x = u + v = 30$$
$$y = u - v = 20$$

6. Cubic equations

The following types occur:

$$x^3 = a \tag{1}$$
$$x^2(x + 1) = a \tag{2}$$
$$x(10 - x)(x + 1) = a \tag{3}$$

The equation (1) was solved by means of a table of cube roots, and (2) by a similar table, as we have seen earlier. We do not know how (3) was solved, but in the case $a = 2, 48$ the Babylonians gave the correct solution $x = 6$.

With $x = 6$, one has

$$x(10 - x)(x + 1) = 6 \times 4 \times 7 = 168 = 2, 48$$

These various examples should be sufficient to show what excellent calculators the ancient Babylonians were in the time of Hammurabi.

Sexagesimal fractions and astronomy

In the time of Hammurabi no astronomical calculations were made, as far as we know. Sumerian arithmetic and Babylonian algebra were developed quite independently of astronomy.

In later times, sexagesimal fractions were also used to express the results of astronomical observations and theories. The zodiacal circle was divided into 12 zodiacal signs of 30 degrees each. Each degree was divided into 60 minutes and each minute into 60 seconds, in accordance with the sexagesimal system. If necessary the division into 60 parts was carried even further.

These divisions are fundamental for all astronomical calculations. The Greeks, Hindus and Arabs continued to use the Babylonian division of the zodiac into 12 signs of 30 degrees each. More generally, they divided every circle into 360 degrees and every degree into 60 minutes, etc. This tradition was continued in geometry and in astronomy up to the present day. The division of the hour into 60 minutes and of the minute into 60 seconds, is also a remnant of the Babylonian sexagesimal system.

The main treatise on Greek astronomy was the so-called *Almagest* of Ptolemy, written about AD 140, whose actual title was *Syntaxis mathematica*, i.e. *Mathematical Composition*. The word 'Almagest' is an Arabic corruption of *Megiste Syntaxis* (*Megiste* = largest) with the Arabic article *Al*.

In the *Almagest*, arcs and angles are always measured in degrees, minutes and seconds. For 47° 42′ 40″ Ptolemy wrote $\mu\zeta\ \mu\beta'\ \mu''$. The diameter of a circle was usually divided into 120 parts, and the parts were again divided according to the sexagesimal system.

For *zero* Ptolemy had a special symbol o (possibly derived from $\omega\delta\epsilon\nu$ = nothing[9]). There was no danger of confusion with the number symbol $o = 70$, because numbers beyond 60 do not occur in counting minutes, seconds, etc.

Ptolemy was a virtuoso in computing with sexagesimal fractions. He multiplied and divided them, he extracted square roots, etc., without wasting a word on the technique of these calculations. So we have to be grateful to his commentator Theon of Alexandria for giving an example of a sexagesimal division.

Theon wanted to divide $\alpha\phi\iota\epsilon\ \kappa'\ \iota\epsilon''$ by $\kappa\epsilon\ \iota\beta'\ \iota''$ or in modern symbols 1515° 20′ 15″ by 25° 12′ 10″. In Theon's calculation, the given figures are to be regarded not as degrees, minutes and seconds of arc, but as pure numbers such as

$$25 + \frac{12}{60} + \frac{10}{60^2}.$$

In contrast to the Babylonians, Ptolemy and Theon used the sexagesimal system only for fractions, not for integers. Thus, 1515 was written in decimal notation.

Theon began by estimating the number of units in the quotient. Our method would be, first to try whether 60 times 25° 12′ 10″ exceeds 1515° 20′ 15″ or not. Theon used a slightly different method. Instead of multiplying 25° 12′ 10″ by 60, he divided 1515° by 60. The quotient 25° 15′ exceeds 25° 12′ 10″. On the other hand, division of 1515° by 61 would give a result less than 25°, hence we have to take 60 as the number of units in the quotient. Next Theon successively subtracted

$$60 \times 25°\ \text{and}\ 60 \times 12'\ \text{and}\ 60 \times 10''$$

from 1515° 20′ 15″. This leaves 190′ 15″.

In the same way, Theon estimated the 7′ of the quotient. Division of 190′ 15″ by 7′ gives more than 25° 12′ 10″, but division by 8′ gives less, so we have to take 7′. The divisions by 7 and by 8 can be done mentally, in a rough approximation, neglecting the 15″. The typical method learnt at school would require the multiplication of 25° 12′ 10″ by 7 and by 8, which is a more complicated operation. Thus, Theon's method has certain advantages.

To divide 1515° 20′ 15″ by 25° 12′ 10″ using Theon's method, one would write:

$$60 \times 25° = \underline{1500°} \qquad\qquad 1515° : 60 = 25° \; 15′$$
$$15° \; 20′ \; 15″ \qquad 1515° : 61 = 24° \; 50′$$

$$60 \times 12′ = \underline{12°} \qquad$$
$$3° \; 20′ \; 15″$$

$$60 \times 10″ = \underline{10′} \qquad$$
$$190′ \; 15″$$

$$7′ \times 25° = \underline{175′} \qquad\qquad 190′ : 7′ > 27$$
$$15′ \; 15″ \qquad\qquad 190′ : 8′ < 24$$

$$7′ \times 12′ = \underline{1′ \; 24″} \qquad$$
$$13′ \; 51″$$

$$7′ \times 10″ = \underline{1″} \qquad$$
$$830″ \qquad\qquad\quad 830″ : 32″ = 26° \text{ nearly}$$
$$830″ : 33″ = 25° \; 9′$$

Hence the quotient is approximately 60° 7′ 33″.

It should be noted that Theon's method of subtracting first 60 × 25°, next 60 × 12′, etc., seems to be adapted to the use of an abacus. The numbers 1500° and 12° and 10′ to be subtracted from 1515° 20′ 15′ were not written down, but just taken away from the pebbles on the abacus.

Fractions in Ancient Greece, China and India

The most ancient reference to fractions in Greek literature is in Homer's *Iliad*:

Two parts of the night are passed, the third remains [Book K, verse 253]

The fractions occurring here are the natural fractions $\frac{2}{3}$ and $\frac{1}{3}$, which also had special names and special notations in ancient Egypt (see p. 131).

In the works of Archimedes (third century BC) there are several calculations with unit fractions and mixed fractions. Thus, in his treatise *On the Measurement of the Circle*, Archimedes proves that the perimeter of the circles lies between $3\frac{10}{71}$ times and $3\frac{1}{7}$ times the diameter. His contemporary Eratosthenes gave $\frac{11}{83}$ times two right angles as an estimate of the obliquity of the ecliptic. Diophantus, in his *Arithmetika* (third century AD), admits as solutions of his problems not only integers but also fractions.

In earlier mathematical treatises, such as the *Elements* of Euclid, only integers were admitted. Instead of saying that one quantity is $\frac{2}{5}$ of another quantity, Euclid would say that their ratio (in Greek: *logos*) is as 2 to 5.

The main reason why fractions were excluded from pure mathematics at the time of Plato seems to be the philosophical conviction that the unit is indivisible. In

Plato's own words, the experts in mathematics were 'absolutely opposed to dividing the unit' (Plato, *Republic*, 525E).

Fractions were scorned and left to merchants, because, so it was said, visible things are divisible, but mathematical units are not.

Of course, merchants and engineers were familiar with expressions like 'two fifths'. Traces of this terminology can be found even in the arithmetical books of the *Elements* of Euclid. In Book 7 there are the following definitions:

Def. 3   A number is part of a number, the less of the larger, when it measures the larger number.
Def. 4   And parts, when it does not measure it.

'Part' means here nth part, where n is an integer. 'Parts' means a number of nth parts, e.g. two fifths. The two definitions presuppose the notions 'unit fraction' and 'mixed fraction' (i.e. fraction of the form m/n).

An application of these definitions is the definition of proportionality:

Def. 20   Numbers are proportional when the first is the same multiple, or the same part, or the same parts of the second that the third is of the fourth.

This definition can be explained by denoting the four numbers by a, b, c and d and considering what the proportionality

$$a : b = c : d$$

means.

If a is a multiple of b, say $a = nb$, it is required that c is the same multiple of d. If a is a part or parts of b, it is required that c is the same part or the same parts of d. Thus, if $a = \dfrac{n}{m}b$, it is required that $c = \dfrac{n}{m}d$.

Book 7 is one of the most archaic books of the *Elements*. The definitions given above are probably due to those Pythagoreans who lived in the middle of the fifth century BC.

The same Pythagoreans also developed a theory of musical chords. According to this theory fundamental musical intervals like the octave, the fifth and the fourth correspond to ratios of simple numbers. The fourth corresponds to the ratio 4:3, and this ratio was called *epitriton*, which means 'on it one third'. In non-scientific Greek speech, the same expression just meant the fraction $1\frac{1}{3}$. Likewise the ratio 9:8, which corresponded to the whole tone according to the Pythagoreans, was called *ep-ogdo-on*, which means 'on it one eighth', or $1 + \frac{1}{8}$. Thus the Pythagoreans started with mixed fractions like $1\frac{1}{3}$ or $1\frac{1}{8}$ familiar from everyday life, and developed a scientific theory in which these fractions were replaced by ratios like 4:3 or 9:8. In their theoretical expositions, only integers were admitted, but the terminology shows that the notion of a mixed fraction $\dfrac{n}{m}$ was in their minds from the very beginning. Archimedes and Eratosthenes were also familiar with this notion.

Archimedes was a virtuoso in handling mixed fractions. In his treatise *On the Measurement of a Circle* he first shows that the perimeter of a circumscribed regular polygon of 96 sides is less than $3\frac{1}{7}$ times the diameter of the circle, and that the perimeter of an inscribed polygon of 96 sides exceeds $3\frac{10}{71}$ times the diameter. From this he concludes that the perimeter of the circle itself lies between the same two limits.

The method to estimate the perimeter of the circumscribed polygon is as follows. Archimedes starts with a circumscribed hexagon and then doubles the number of sides again and again. The side of the hexagon can be calculated from a rectangular triangle having an angle of 30° at the vertex E (see Figure 88). If the side e opposite

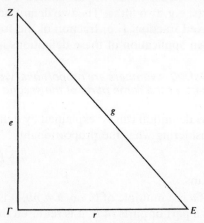

FIGURE 88   Part of Archimedes' method for estimating the perimeter of a circumscribed polygon

E is taken as the unit of length, the hypotenuse g is 2,[10] and the square of the third side r (the radius of the circle) is, according to the Theorem of Pythagoras,

$$r^2 = g^2 - e^2 = 4 - 1 = 3$$

Hence the ratio of r to e is what we would call $\sqrt{3}$. Archimedes now shows that this ratio is contained between narrow limits

$$\frac{265}{153} < \frac{r}{e} < \frac{1351}{780}$$

Archimedes does not say how he obtained these results, but it is easy to verify his inequalities, for the square of $\frac{265}{153}$ is less than 3, and the square of $\frac{1351}{780}$ is more than 3.

It is interesting to see just how accurate Archimedes' limits are. The value of $\sqrt{3}$ to 7 decimal places is 1.7320508. The lower limit is 1.7320261 and is accurate to 4 decimal places. The upper limit is 1.7320513 and is accurate to 6 decimal places.

In Greek, the fraction $\frac{1}{3}$ was called *to triton*. The word *triton* could be written out in full, or it might be abbreviated to $\gamma^{\rho\nu}$ or still shorter to γ' or γ''. These notations are known from the works of Greek mathematicians like Archimedes and Diophantus, and also from preserved papyri.

The papyri are a more reliable source, because the works of the mathematicians were copied again and again, and the copyists might well have changed the notations. Two examples of these papyri are the Papyrus Vindobonesis 19 996, published by Gerstinger and Vogel in *Mitteilungen Papyrussammlung Erzherzog Rainer*, Vienna 1932, and the Papyrus Akhmim, published by J. Baillet in *Memoires de la Mission archéologique française au Caire 9*, fasc. 1, Paris 1892. In the latter papyrus mixed fractions are expressed as sums of unit fractions, for example:

The seventeenth part of 3 is $\dfrac{1}{12} + \dfrac{1}{17} + \dfrac{1}{51} + \dfrac{1}{68}$.

The Papyrus Akhmim was written in Greek, in Hellenistic Egypt. It seems that the scribe who wrote it was familiar with ancient Egyptian methods of calculation. If he used the (2:*n*) table of the Rhind Papyrus, he could perform the multiplication $3 \times \overline{17}$ as follows:

$$
\begin{array}{ll}
/\ 1 & \overline{17} \\[4pt]
/\ 2 & \overline{12} + \overline{51} + \overline{68} \\[4pt]
\text{Total} & \overline{12} + \overline{17} + \overline{51} + \overline{68}
\end{array}
$$

The Greeks were also able to perform the opposite operation, namely to convert a sum of unit fractions into a mixed fraction. Thus, Nikolaos Rhabdas, who lived in Byzantium in the fourteenth century AD, adds $3 + \dfrac{1}{3} + \dfrac{1}{14} + \dfrac{1}{42}$ as follows:

$$\frac{1}{3} = \frac{14}{42}$$

$$\frac{1}{14} = \frac{3}{42}$$

$$\frac{1}{42} = \frac{1}{42}$$

$$\text{Total} \qquad \frac{18}{42} = \frac{3}{7}$$

$$\text{Add} \qquad 3 = \frac{21}{7}$$

$$\text{Total} \qquad \frac{24}{7}$$

For mixed fractions like $\dfrac{3}{5}$, the Greeks had two contradictory notations. Frequently the denominator was placed above the numerator:

$$\frac{3}{5} = \frac{\epsilon}{\gamma}$$

According to Vogel[11] this notation came into use in the time of Archimedes (third century BC). However, the Papyrus Vindobonus (first century AD) has just the opposite notation

$$\frac{3}{5} = \frac{\gamma}{\epsilon}$$

In ancient Hindu texts on calculation, the same notation was used as in the Papyrus Vindobonus. It seems that our notation $\frac{n}{m}$ was derived either from the Greek or from the Indian notation.

The following two expressions can be convertd into mixed fractions, using the Greek method:

$$\text{(a)} \quad 2 + \frac{1}{6} + \frac{1}{14} + \frac{1}{21}$$

$$\text{(b)} \quad 4 + \frac{1}{4} + \frac{1}{5} + \frac{1}{6}$$

(a)

$$\frac{1}{6} = \frac{7}{42}$$

$$\frac{1}{14} = \frac{3}{42}$$

$$\frac{1}{21} = \frac{2}{42}$$

Total $\quad \frac{12}{42} = \frac{2}{7}$

Add $\quad 2 = \frac{14}{7}$

Total $\quad \frac{16}{7}$

(b)

$$\frac{1}{14} = \frac{15}{60}$$

$$\frac{1}{5} = \frac{12}{60}$$

$$\frac{1}{6} = \frac{10}{60}$$

Total $\quad \frac{37}{60}$

Add $\quad 4 = \frac{240}{60}$

Total $\quad \frac{277}{60}$

In *Science and Civilization in China*, Needham quotes from a Chinese author of the first century BC who gives rules for calculating with fractions. Of Chinese sources in general Needham writes:

In 1936 Têng Yen-Lin and Li Nien published a union catalogue of Chinese mathematical books in the libraries of Peking which contains somewhat over 1000 titles, though some of them are European books translated either by the 17th-century Jesuits or by later foreign scholars such as Wylie, Edkins and Mateer. The bulk of the material, however, is prior to the European influence, or by Chinese mathematicians working largely independently during the Ching dynasty[12].

. . . .

A high proportion of the earlier books on mathematics, before the Sung (+ 13th century), are irretrievably lost – we know them by the titles which appear in the bibliographies of the official histories, and by references in other writings.

The first-century work which treats of fractions is the *Chiu Chang Suan Shu* (the *Nine Chapters on the Mathematical Art*). In the first of the *Nine Chapters* the rules of calculation with fractions are explained. The author has no special notation for fractions like our notation $\frac{3}{5}$: he just says 'of 5 parts 3'. The denominator was called *mu*, which means *mother*, and the numerator *tzu*, which means *son*.

For the simplification of fractions the following rule was given:

If [denominator and numerator] can be halved, do it. If not, lay out [on the Abacus] the denominator and numerator [as numbers of pebbles]. The smaller [number] subtract from the larger one. Thus, change the numbers, diminishing them by alternate subtractions, until you obtain equal numbers. Divide [denominator and numerator] by this equal number.

This procedure is equivalent to the well-known 'Euclidean algorithm'[13] for finding the largest common divisor of two numbers by successive divisions. An example is worked below, the object being to find the largest common divisor of 10 and 4, first by the Chinese method, and then by the usual method of successive divisions.

Chinese method:

1st step $10 - 4 = 6$

2nd step $6 - 4 = 2$ ⎫
3rd step $4 - 2 = 2$ ⎬ Now the numbers are equal
⎭

Usual method:

1st step Divide 10 by 4; remainder 2

2nd step Divide 4 by 2; remainder 0

Both cases give (of course) the number 2 as the lowest common divisor. The Chinese method involves subtractions, the usual method involves divisions.

Addition of fractions was performed by taking the product of all denominators as a new denominator for all fractions, and then adding all numerators, thus:

$$\frac{a}{b} + \frac{c}{d} + \frac{e}{f} = \frac{adf + cbf + ebd}{bdf}$$

Subtraction of fractions was performed as follows:

$$\frac{a}{b} - \frac{c}{d} = \frac{ad - cb}{bd}$$

Multiplication was performed by multiplying denominators and numerators:

$$\frac{a}{b} \cdot \frac{c}{d} = \frac{ac}{bd}$$

A division of fractions was performed thus: first the two fractions were reduced to the same denominator, next the numerators were divided:

$$\frac{a}{c} : \frac{b}{c} = \frac{a}{c}$$

Comparing the addition of the fractions $\frac{1}{30}$ and $\frac{1}{42}$ first by the typical 'school' method and secondly by the Chinese method shows which of the two methods is the simpler. The 'school' method would reduce both fractions to the common denominator 210:

$$\frac{1}{30} + \frac{1}{42} = \frac{7}{210} + \frac{5}{210} = \frac{12}{210} = \frac{2 \times 6}{35 \times 6} = \frac{2}{35}$$

The Chinese method would yield

$$\frac{1}{30} + \frac{1}{42} = \frac{42}{30 \times 42} + \frac{30}{30 \times 42} = \frac{72}{1260}$$

Since the numerator and denominator are both even, they may be halved and again halved:

$$\frac{72}{1260} = \frac{36}{630} = \frac{18}{315}$$

Next one has to determine the largest common multiple of the numerator and denominator by successive subtractions. Of course the result is 9. Now the numerator and denominator are divided by 9:

$$\frac{18}{315} = \frac{2}{35}$$

Obviously, the 'school' method is much simpler.

It is very instructive to compare the Chinese rules with the corresponding rules given in Hindu treatises. Brâhmagupṭa, who wrote his astronomical treatise *Brâhmasphuṭa-siddhânta* in AD 628, gives the rule for adding or subtracting two fractions as follows:

By the multiplication of the numerator and denominator by each of the [fractional] quantities by the other denominators, the quantities are reduced to a common denominator. In addition, the numerators are united [i.e. added]. In subtraction their difference is taken.[14]

Brâhmagupta formulates the rule for addition only for the case of two fractions. A later author, Srîdhara, in AD 750 generalises the rule as follows:

> To reduce to a common denominator, multiply the numerator and denominator of each fraction by the other denominators.[15]

This is just the same as the Chinese rule for addition of fractions.
Brâhmagupta's rule for multiplication reads:

> The product of the numerators divided by the product of the denominators is the [result of] multiplication of two or more fractions.[16]

This rule too is identical with the Chinese rule. A later author, Mahâvîra (ninth century AD), refers to cross-reduction:

> In the multiplication of fractions, the numerators are to be multiplied by the numerators and the denominators by the denominators, after carrying out the process of crosswise cancellation, if that be possible.[17]

Brâhmagupta's method of division is exactly the same as the 'school' method.

> The denominator and numerator of the divisor having been interchanged, the denominator of the dividend is multiplied by the [new] denominator and its numerator by the [new] numerator. Thus, division of fractions is performed.[18]

Earlier authors such as Âryabhaṭa and Brâhmagupta have no preference for unit fractions, but Mahâvîra gives some rules for expressing arbitrary fractions as sums of unit fractions. The Egyptians and Greeks too had methods to express arbitrary fractions as sums of unit fractions. It is quite possible that Mahâvîra was influenced by Greek sources. Three examples of these methods are shown below.

(a) Srîdhara's method to obtain the sum:

$$\frac{4}{5} + \frac{3}{10} + \frac{11}{70}$$

(b) Mahâvîra's method to obtain the product:

$$\frac{3}{4} \times \frac{12}{35} \times \frac{7}{10}$$

(c) Brâhmagupta's method to obtain the quotient:

$$\frac{6}{35} \div \frac{8}{21}$$

These give:

$$\text{(a)} \quad \frac{4}{5} + \frac{3}{10} + \frac{11}{70}$$

$$= \frac{2800}{3500} + \frac{1050}{3500} + \frac{550}{3500}$$

$$= \frac{4400}{3500}$$

$$= \frac{44}{35}$$

$$\text{(b)} \quad \frac{3}{4} \times \frac{12}{35} \times \frac{7}{10}$$

$$= \frac{3}{1} \times \frac{3}{5} \times \frac{1}{10}$$

$$= \frac{9}{50}$$

$$\text{(c)} \quad \frac{6}{35} \div \frac{8}{21}$$

$$= \frac{6}{35} = \frac{21}{8}$$

$$= \frac{126}{280}$$

$$\left[= \frac{9}{20} \right]$$

The fraction $\frac{n}{m}$ was written as $\begin{smallmatrix}n\\m\end{smallmatrix}$ in Greece as well as in India. The Greek notation is known from a papyrus dating from the first century AD. According to Datta and Singh[19], the Hindu notation came into use about AD 200.

Now the question arises: Did the Hindu mathematicians invent the notation anew, or did they learn it from the Greeks? In order to investigate this question, it is necessary to ask first whether there are other traces of Greek influence in the works of Hindu mathematicians.

The earliest datable book of a Hindu mathematician and astronomer is the *Âryabhatîya* of Âryabhata (translated by W. E. Clark, Chicago, 1930). The author, sometimes called Âryabhata I, states that he was 23 years in 499 AD, when just 3600 years of *Kaliyuga* or Iron Age had passed. The doctrine of the Four Ages (gold, silver, etc.) is also known from Greek sources. Âryabhata's book consists of four parts, of which one is called *Ganita*, i.e. Calculation; the other three parts deal with astronomy. Âryabhata was an excellent astronomer; his astronomical system was based on very accurate observations made around AD 500.

Âryabhata's astronomy was based upon the Greek theory of epicycles and excenters. His main mathematical tool was trigonometry, which was invented by the Greeks. Just like the Babylonians and Greeks, Âryabhaṭa divided the zodiacal circle into 12 zodiacal signs of 30 degrees each, and he also divided the degree into 60 minutes, and so on.

He sometimes used loan-words from Greek, such as *kendra* (from Greek *kentron* = centre) and *lipta* (from Greek *lepta*, plural of *lepton* = minute). Thus we find several traces of Greek influence in the work of Âryabhaṭa.

Varâha Mihira, who lived in the middle of the fifth century, one or two generations after Âryabhaṭa, is the author of an astronomical compendium called *Panča-siddhântikâ*. *Panča* means 'five', and *Siddhânta* means 'textbook of astronomy'. In fact, the *Panča-siddhântikâ* contains summaries of five astronomical textbooks existing at the time of Varâha Mihira. Two of these five Siddhântas are based upon Babylonian astronomy, while the three others, *Sûrya-Siddhânta, Romaka-Siddhânta* and *Puliśa-Siddhânta* are based upon Greek astronomy and trigonometry. Varâha Mihira tells us that these three Siddhântas are more accurate than the other two. The *Sûrya-Siddhânta* is closely related to the work of Âryabhaṭa, and the *Romaka-Siddhânta* (*Romaka* means the Roman empire, including Greece and Hellenistic Egypt) is just as closely related to the work of the great Greek astronomer Hipparchus, who lived about 130 BC. The early Hindu astronomers such as Âryabhaṭa and Varâha Mihira were obviously well acquainted with Greek astronomical theories.

To return to arithmetic, Âryabhaṭa's chapter on *gaṇita* (calculation) is sandwiched between astronomical chapters. Moreover, calculations are a necessary tool in astronomy. It is not possible to learn Greek astronomy without learning Greek methods of calculation. It is therefore only natural to suppose that Âryabhaṭa was also familiar with Greek methods of calculation.

Now the fact that the Hindu notation for fractions was just the same as the Greek notation can be considered as an additional argument in favour of the assumption that the Hindus learnt arithmetic (directly or indirectly) from the Greeks.

Âryabhaṭa describes the method of finding the square root of an integer very concisely:

> Always divide the even place by twice the square-root; after having subtracted from the odd place the square, the quotient put down in the next place gives the root.[20]

This description is so short that it can hardly be understood without commentary. Later writers give more details of the process. Thus Śrîdhara says:

> Having subtracted the square from the odd place, divide the next [even] place by twice the root which has been separately placed, and after having subtracted the square of the quotient, write it down in the line. Double what has been obtained above and taking this down, divide by it the next even place. Halve the doubled quantity.

This process can be illustrated by finding the square root of 54756. (The odd and even places, counting from the right, will be marked by vertical and horizontal strokes.)

$$| \ - \ | \ - \ |$$

	5	4	7	5	6	

Subtract square   4   root = 2

Divide by twice the root   4)   14   (3   quotient = 3

                                    12

                                    27   new root = 23

Subtract square of quotient   9

Divide by twice the root   46)   185   (4 quotient = 4
                                        184

                                        16   final root = 234

Subtract square of quotient   16

                                            0

What is the idea underlying this calculation? Let D be the given number and let a be an approximate value of the square root. In this case

$$D = \ 5 \ \ 47 \ \ 56 \quad \text{and} \quad a = 400$$

The following equation now has to be solved:

$$(a + x)^2 = D$$

or

$$2ax + x^2 = D - a^2$$

Since x^2 is small as compared with $2ax$, one may, in a rough approximation, neglect the term x^2. Solving for x, one obtains an approximation x_1, which Srîdhara consistently calls the quotient:

$$x_1 = \frac{D - a^2}{2a}$$

Only the first decimal place of x_1 is taken into account. Now the second approximation to the required square root will be $a + x_1$.

To continue the process, one subtracts

$$(a + x_1)^2 = a^2 + 2ax_1 + x_1^2$$

from D. Having already subtracted a^2, one has to subtract $2ax_1$ and x_1^2. This is just what Srîdhara prescribes.

The fundamental formula on which the whole method is based is

$$(a + x)^2 = a^2 + 2ax + x^2 \tag{1}$$

Srîdhara and his predecessors Âryabhata and Brâhmagupta certainly knew this formula, for their method of calculating squares is based upon it.[21] Mahâvîra also gives (in words, of course) the more general formula

$$(a + b + c + \cdots)^2 = a^2 + b^2 + c^2 + \cdots + 2ab + \cdots \tag{2}$$

Brâhmagupṭa and Śrîdhara also make use of the formula[22]

$$(a + b)(a - b) = a^2 - b^2 \tag{3}$$

The formulae (1), (2) and (3) were also known to the Babylonians and to the Greeks. In the *Elements* of Euclid, (1) and (3) appear in geometric form as theorems 4 and 5 of Book 2. Probably the Hindu mathematicians learnt the use of formulae (1), (2) and (3) from the Greeks.

Another sign of Greek influence is the formula

$$n^2 = 1 + 3 + 5 + \cdots \quad \text{to } n \text{ terms}$$

which was mentioned by Śrîdhara and Mahâvîra.[23] This formula also occurs in Greek sources such as the *Arithmetical Introduction* of Nicomachus of Gerasa.[24] This *Introduction*, probably written in the first century AD, enjoyed great popularity up to the end of the Middle Ages. It presented a very agreeable mixture of popular philosophy, number mysticism and arithmetic.

This section concludes with a solution to a quadratic equation given by Âryabhaṭa. The problem is one of compound interest which gives rise to a quadratic equation

$$tx^2 + px = Ap$$

The principal sum p (= 100) is lent for one month at unknown interest x. This unknown interest is then lent out for t (= 6) months. After this period, the original unknown interest plus the additional interest amounts to A (= 16). The (positive) solution is given as

$$x = \frac{\sqrt{Apt + \left(\frac{p}{2}\right)^2} - \frac{p}{2}}{t}$$

In Âryabhaṭa's own words:

> Multiply the sum of the interest and the interest on this interest [A] by the time [t] and by the principal [p]. Add to this result the square of half the principal $\left[\left(\frac{p}{2}\right)^2\right]$. Take the square root of this. Subtract half the principal $\left[\frac{p}{2}\right]$ and divide the remainder by the time [t]. The result will be the [unknown] interest [x] on the principal.[25]

As previously stated, the solution of quadratic equations was known to the Babylonians, and the Greeks too were familiar with both geometrical and arithmetical forms of the solution.

The main points of contact between Babylonian, Greek and Hindu arithmetic and algebra can be summarised as follows: the formula $(a + x)^2 = a^2 + 2ax + x^2$ and its generalisation to sums of more than two terms; the methods of extracting square roots based on this formula and on the idea of neglecting the x^2 term; the formula $(a + b)(a - b) = a^2 - b^2$; the expression of a square n^2 as a sum of successive odd numbers; the solution of quadratic equations. Thus, Hindu arithmetic and algebra are probably derived from Babylonian and Greek traditions.

In this chapter fractions and calculations involving notations sometimes very different from our own have been considered, for example the writing of mixed fractions as sums of unit fractions, sexagesimal fractions, and so on. Many of the calculation methods examined have been unfamiliar. Though they can be related to our own methods in some cases (for example the Egyptian auxiliary numbers, the Babylonian methods for solving equations, and Hindu methods of manipulating mixed fractions), in other cases they are very different (for example, the Egyptian use of $(2:n)$ tables, the Babylonian use of reciprocals, and the Hindu extraction of square roots).

It should be remembered, however, these calculations were undertaken by the scholars – priests, scribes and astronomers. Since the great majority of the people could neither read nor write, recourse had to be made to the counting board (or abacus) for the everyday needs in commerce of the relatively uneducated. The history of the counting board is considered in Chapter 6.

Decimal fractions in China and among the Arabs

Decimal fractions were invented three times: first in China in the third century AD; then in Western Europe in the fourteenth century; and once again by the Muslim astronomer Jemshid al-Kashi in the fifteenth century. They came into general use at the end of the sixteenth century, through the work of Simon Stevin of Bruges. The three discoveries were apparently independent.

The emperor Chhin Shih Huang Ti, who unified the Chinese empire in 221 BC, standardised the units of length. The *chhih*, a small foot of about 24 cm, was divided into 10 *tshun*, and then further divided as follows:

$$1 \; tshun = 10 \; fên$$
$$1 \; fên \quad = 10 \; li$$
$$1 \; li \quad \; = 10 \; fa$$
$$1 \; fa \quad \; = 10 \; hao$$

Of course, extremely small units like the *fa* and the *hao* were never used in actual measurements, but for the purposes of calculation this decimal division was very convenient. Thus, the mathematician Liu Hui, who lived in the third century AD, expressed a diameter of 1.355 feet as

1 chhih, 3 tshun, 5 fen, 5 li

Liu Hui used the same system to express pure numbers as decimal fractions. Commenting on the extraction of square roots, he said that their fractional parts may be expressed by taking the denominator, first as 10, then as 100, and so on as far as one wants.

After the time of Liu Hui there was little change in the methods. The book *Sui Shu*, dated AD 635, expressed the length of the circumference of a circle, diameter 1 *chang* (= 10 *chhih*), as follows:

3 chang 1 chhih 4 tshun 1 fên 5 li 9 hao 2 miao 7 hu

In other words the author used the excellent approximation 3.1415927 for what we now call π. The approximation is exact up to the last decimal!

A later author Han Yan (about AD 800) seems to have been the first to drop the names of the powers of ten and to write down the numbers as in our system, using a word instead of the decimal point to indicate the last integer place.

From the eighth century on, Samarkand flourished as a trade centre on the route between Baghdad and China (see Figure 89). In the ninth and tenth centuries, the town became a centre of Islamic civilisation, and later, in the fourteenth century, it was the capital of Tamerlane's empire. His grandson Ulugh-Beg established an astronomical observatory at Samarkand in 1420.

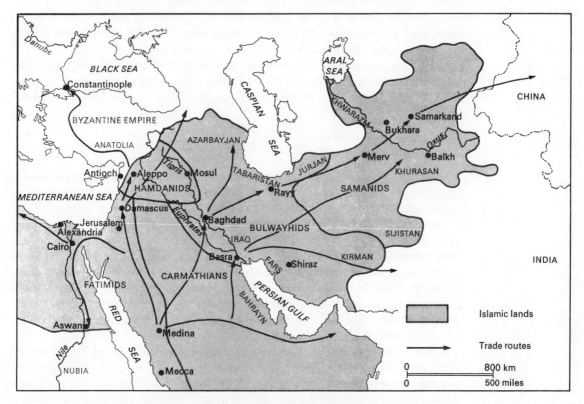

FIGURE 89   Islamic trade routes in the Middle Ages

One of the leading mathematicians working at this observatory was Jemshid al-Kashi, who died in 1436. He was a very skilful calculator. (In his *Treatise on the Circumference* he calculated π to 16 decimal places[26].) In another book called *The Key to Arithmetics* al-Kashi first described calculations with sexagesimal fractions, and then said that by analogy he had introduced fractions composed of powers of $\frac{1}{10}$. His aim, according to his own words, was to create a system in which all operations could be performed in the same way as with integers.

To separate integers from fractions, al-Kashi had several methods. Sometimes he used a vertical bar, or he wrote the fractional part in another colour. In most cases he just wrote the ordinal number of the last decimal place above the last figure.

Thus, 25 would mean 0.25. From Byzantine sources we know that al-Kashi's decimal fractions were used in Turkey in the fifteenth century.[27]

Decimal fractions in Western Europe

In the Middle Ages, the development of decimal fractions was mainly due to the astronomers. Two factors were of decisive importance: first the existence of sexagesimal fractions, well known from 'minutes' and 'seconds', and secondly the need of the astronomers to simplify trigonometric calculations.

About 1350 the Rabbi Immanuel Bonfils of Tarascon in southern France proposed the division of unity into ten parts, which he called 'Primes', and the further division of each Prime into ten parts, called 'Seconds', and so on. In the word 'Second' can be seen the analogy with the well-known sexagesimal division of the degree and the hour. Bonfils gave some rules for multiplication and division, but no examples. It seems that his system was soon forgotten.

About the same time the Parisian astronomer Jean de Meurs (also called Johannes de Muris), computed $\sqrt{2}$ as one thousandth of the square root of 2 millions. This square root, rounded to the nearest integer, is 1414, so he could write his result as

$$\sqrt{2} = \frac{1}{1000} \cdot 1414$$

John of Meurs commented that the result could also be expressed by saying that the square root of 2 was 1414, the first digit being taken as an integer, the next as a tenth, etc.

The same tendency to multiply fractions by a power of ten and to round the result to the nearest integer can also be observed in trigonometry. In medieval tables of sines the angles were always measured in degrees, minutes and seconds, but instead of our sin x, the function $R \sin x$ was tabulated, R being an arbitrarily chosen radius. Ptolemy, in his table of chords, had chosen $R = 60$, but the fifteenth-century Viennese astronomer Georg Peurbach chose R as 60 000, which enabled him to avoid fractions. His table of $R \sin x$ nearly reached the accuracy of a modern five-place sine table.

Peurbach's pupil, Johann Mueller of Königsberg (usually called Regiomontanus from *Rex = König* = 'king' and *Montanus = von Berg* = 'belonging to a mountain'), increased the accuracy by choosing $R = 6\,000\,000$. In his tangent tables Regiomontanus took a further step in the direction of the decimal system: he first chose $R = 10^5$ and next $R = 10^7$, thus avoiding the divisions by 60 that were needed in the earlier system. A tangent table based on $R = 10^5$ is equivalent to a modern five-place decimal tangent table: we need only insert the decimal point. Later tables such as the *Opus Palatinum* were all based on a radius equal to a power of 10. Thus, at the beginning of the sixteenth century, the scene was well prepared for the introduction of decimal fractions.

In 1530, Christoff Rudolff wrote an 'Example Booklet', in which he explained the compound interest calculus. He calculated a table for

$$375\left(1 + \frac{5}{100}\right)^{n}$$

the exponent n ranging from 1 to 10. He wrote the results in a notation which only differs from present notation in the use of a vertical dash instead of the decimal point. Thus, he wrote the result for $n = 2$ as 413|4375.

One of the pioneers of algebra was the Frenchman François Viète, usually called Vieta. In his *Canon mathematicus* of 1579 the occasional decimal fraction is to be found written without a denominator. The fractional part was printed in smaller type and underlined. Thus, assuming the radius of a circle to be 100,000<u>000.00</u>, Vieta showed that the semi-perimeter lay between 314,159<u>265,35</u> and 314,159<u>265,37</u>. He used commas to arrange the digits in groups of threes. In the same book there is also for $R \sin 60°$ the value 86,602|540,37, the radius R being taken as 100,000.

Although Simon Stevin had several predecessors in Western Europe, until his work was published decimal fractions were introduced only casually in limited domains of application. Simon Stevin was the first to give a systematic exposition of the whole system.

Simon Stevin of Bruges spent most of his life in Holland. His pamphlet *De Thiende* ('the tenth') was printed in 1585. A French translation was published in the same year under the title *La Disme* (see Figure 90). The Dutch treatise was reprinted in 1626 and again in 1630. Two English translations, one by Robert Norton and a freer one by Henry Lyte, appeared in 1608 and 1619 respectively. The 'Tenth' (or 'Dime') became Stevin's most popular book.[28]

In his Preface, Stevin asked whether the method explained in his book could be described as an 'admirable invention'. His answer was negative; he said it was more like the chance discovery of a treasure. He only claimed that the method was extremely useful, especially for star gazers, surveyors, measurers of tapestry and merchants.

In the first part following the Preface, Stevin explained his notation. The integral part of any number was called its *commencement*, and given the sign ⓪. Thus, if the integer was 364, he wrote 364 ⓪. The tenth part of a unit was called *first* or

FIGURE 90   Extract from Stevin, *La Disme* (Turner Collection, University of Keele)

prime, and was represented as ①. The tenth part of the Prime was called *second*; its sign was ②, and so on. Thus, 3①7②5③9④ meant 0.3759.

In the second part of his work Stevin explained the addition, subtraction, multiplication and division of decimal numbers. His first example of an addition looks like this:

$$
\begin{array}{ccccc}
⓪ & ① & ② & ③ & \\
2 & 7 & 8 & 4 & 7 \\
3 & 7 & 6 & 7 & 5 \\
8 & 7 & 5 & 7 & 8 & 2 \\
\hline
9 & 4 & 1 & 3 & 0 & 4 \\
\end{array}
$$

Stevin wrote the result as 941⓪3①0②4③ and proved that the sum was correct.

The multiplication 32.57 × 89.4 was done as follows:

$$
\begin{array}{ccccccc}
 & & ⓪ & ① & ② & \\
 & & 3 & 2 & 5 & 7 \\
 & & 8 & 9 & 4 & 6 \\
\hline
 & & 1 & 9 & 5 & 4 & 2 \\
 & 1 & 3 & 0 & 2 & 8 \\
 2 & 9 & 3 & 1 & 3 \\
2 & 6 & 0 & 5 & 6 \\
\hline
2 & 9 & 1 & 3 & 7 & 1 & 2 & 2 \\
 & & ⓪ & ① & ② & ③ & ④ \\
\end{array}
$$

Divisions were performed in the usual way. At the end of his Second Part, Stevin also showed how to extract a square root.

Stevin's method rapidly spread to England, France, Germany and Italy. His cumbersome notation ⓪ ① ② ... was retained by some Dutch and French authors, but others soon simplified it.

The first to simplify Stevin's notation seems to have been the astronomer Magini of Bologna. In his text on plane triangles (1592) he used decimal fractions such as 6822,11, separating the integral and fractional parts by a comma.

Clavius, the influential Vatican astronomer and textbook writer, included in his *Astrolabium* of 1593 a table of the function $R \sin x$ based on $R = 10^7$. The function values were given as integers but, in the little tables of *partes proportionales*, entries such as 34.4 appear.

Jost Bürgi, a Swiss instrument maker and one of the inventors of logarithms, wrote a manuscript *Arithmetica*, some time after 1592, in which 141.4 was written as 141ₒ4. The manuscript was not printed. On the other hand, on the title page of Burgi's *Progress Tabulen*, printed in 1620, there is the notation

$$230270 \cdot 022$$

The first printed table of Logarithms was published by John Napier, Laird of Merchiston, in 1614. The title of the book was *Mirifici Logarithmorum canonis descriptio*. In this book no decimal fractions occur: all sines and logarithms in Napier's tables were multiplied by 10^7 so as to obtain integers. However, they do occur in a posthumous treatise of Napier, published in 1619 under the title *Mirifici Logarithmorum canonis constructio*. Napier used the same notation as Clavius, writing 25.803 just as most people do nowadays. Right at the beginning of the treatise he stated: 'Whatever is written after the period is a fraction.'

The great astronomer, Johannes Kepler (1571–1630) considered Bürgi as the inventor of decimal fractions. He wrote 3(65 instead of Bürgi's 3_o65.

Johann Hartmann Beyer of Frankfurt-on-the-Main claimed to have invented the calculus of decimal fractions in 1597 under the influence of 'star-artisans', i.e. astronomers. In his *Logistica decimalis* of 1619 is such notation as

	o	I	II	III	IV	V	VI
1	2	3	4	5	9	3	7	2

for 123.459372. It is not known whether Beyer's invention was really independent of Simon Stevin's influence.

Briggs, who continued the work of Napier and computed a table of logarithms with basis 10 in 1624, used a comma to separate integers from fractions. The Dutchman Adriaen Vlacq, who completed Briggs's table, continued to use the comma. His tables of logarithms were published in 1627. From then on the use of decimal fractions spread rapidly over Western Europe.

For purposes of comparison, the number 3.14 is written below in the notations of Stevin, Magini, Clavius, Bürgi, Kepler, Beyer, Napier, Briggs and Vlacq:

Stevin	3 ⓪ 1 ① 4 ②
Magini	3,14
Clavius	3.14
Bürgi	3_o14 or 3·14
Kepler	3(14
Beyer	o I II 3 1 4
Napier	3.14
Briggs	3,14
Vlacq	3,14

After the introduction of trigonometry, decimal fractions and logarithms, numerical calculations no longer presented any problems. The stage was now set for the appearance of the great mathematicians of the seventeenth century: Descartes, Fermat, Pascal, Libniz, Huygens and the greatest of all, Newton.

6 Aids to calculation

The abacus in antiquity

In a positional system like our own, calculations with pencil or pen and paper are easy, but in a non-positional notation like that of the Romans they are much more difficult. To see this more clearly, consider the multiplication of 325 by 47. Using our numerals, the calculation is simple:

$$
\begin{array}{r}
325 \\
47 \\
\hline
2275 \\
1300 \\
\hline
15275
\end{array}
$$

Now try to multiply CCCXXV by XLVII. In the former system, one had to multiply every single figure contained in 325 with every figure in 47 (altogether 6 simple multiplications), to write the partial results in the right places, and to add them. If one tries to do the same thing with CCCXXV and XLVII, the first problem that arises is that XLVII cannot be decomposed into parts X + L + V + I + I, because the notation XL is subtractive. One can try writing XXXX instead of XL and try computing the product CCCXXV.XXXXVII by a method similar to today's, multiplying every single component C or X or V contained in the first factor by every component X or V or I of the second factor. This method, however, will involve 42 single multiplications, followed by the addition of the results. A very cumbersome method!

Trying to find the partial product CCC.XXXX in one single operation involves three times C and four times X, which is twelve times the product C.X, or twelve thousand, written as

ↀMM

Proceeding in this way, the final result

ↀↁ CCLXXV

is obtained by adding only six partial products. Yet the whole procedure is still very cumbersome.

Now the question arises: How did the Romans themselves perform such calculations? The answer is: they used an *abacus* or *counting board* to assist them.

On a counting board, numbers are marked by pebbles, also called *counters*. The board is divided into columns or strips on which the pebbles are laid. The basic form of such a counting board is shown in Figure 91. Each pebble in the first

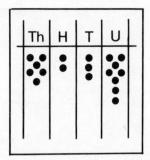

FIGURE 91 Basic form of the counting board, showing the number 6238

column has the value *thousand*, in the second column *hundred,* and so on. The number shown is 6238.

In order to calculate the product 325 × 47 on the abacus, both factors first have to be displayed, one below the other. Next the six partial products 3 × 4, 3 × 7 have to be formed by mental multiplication, and so on. These have to be laid out at the appropriate places on the counting board. Ten units may be replaced by one unit in the column of tens, etc. Thus the desired product 15275 is obtained by very easy operations, which may afterwards be written down as

ⅭⅠƆƆ CCLXXV

From this description it can be seen that the abacus has the same advantages as a place-value system. In fact, laying out a number on the abacus is practically equivalent to writing it down in place-value notation.

There is no indication that the abacus was known in ancient Egypt or in Babylonia. However, there is proof of its existence in classical antiquity: a few counting boards actually used by the Greeks and Romans have been preserved and so have contemporary portrayals of them.

The only ancient Greek counting board that has been preserved is a tablet of white marble found about the middle of the last century on the island of Salamis (see Figure 92). Its exact date is not known.

Two groups of parallel lines have been chiselled into this tablet, one group of eleven lines crossed by a perpendicular line through their middle, and another group of five shorter ones at some distance from the first group. Along the two long sides of the tablet and across one of the shorter sides are letters, which can be identified as early Greek numerals and denominations of ancient coins:

T ᚠX ᛗH ᛗΔ ᚠⱵ ⅠCTX

Talent =6000	1000	100	10	1	1½ ¼ ⅛
		Drachmas			Obol

The three symbols at the right are those for the half, the quarter and the eighth obol, which were respectively called the *hemiobolion* ('half-obol') Ϲ, the *tetartemorion* T ('fourth part') and the *chalkos* X ('bronze'). The last two are not to be confused with the symbols X for *chilioi* and T for *talanton*, which stand at the left end of the row of numerals. The Salamis Tablet was probably used for calculations in a government financial office.

Looking at the tablet, one can well understand a passage written by the historian

FIGURE 92   Salamis tablet (TAP Service.
Athens)

Polybius (second century BC):

> The courtiers who surround kings are exactly like counters on the lines of a
> counting board, for, depending on the will of the reckoner, they may be valued
> either at no more than a mere chalkos, or else at a whole talent.

The chalkos and the talent, of course, are the lowest and the highest values on the
tablet, at the extreme right and left ends of the scale. No information has been
preserved concerning the actual operations performed on this abacus; nevertheless
we can infer these with a high degree of probability from the general arrangement
of the Salamis Tablet, from the remaining fragments of other Greek counting
boards with columns drawn or incised on them, and above all from the medieval
counting boards, which we know much more about.

According to Menninger the Greeks used the Salamis Tablet as follows (see
Figure 93):

> To use the Salamis Tablet the reckoner would stand facing it on the right-hand long
> side. His *psephoi* (counters) would be piled in the middle between the two groups
> of parallel lines. The group of lines at the left end of the board as he faced it were
> used for the integers, the group at the right end for fractions. The small crosses
> indicated the columns: the one at the right distinguished the units and the one at
> the left marked off the talents, or the myriads if the computation did not involve
> money.

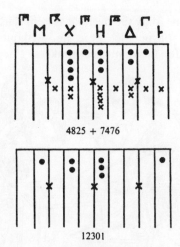

4825 + 7476

12301

FIGURE 93   4825 + 7476 = 12301
on the Salamis Tablet

Let us say, for example, that the quantity 7476 is to be added to 4825. First the reckoner forms the number 4825 with pebbles in the appropriate columns, thereby observing the rule of quinary grouping which is always followed in each even column (the 2nd, 4th, etc), as shown. Then he simply 'gives' the second number 7476 (indicated by x) to the first, as expressed precisely by the Greek technical term *syntithenai* and the Latin *addere* (from *ad-dare*), which became our word 'addition'. The first step of this procedure thus consists of 'writing' or putting down the initial number, the second step of addition proper, and the third step of grouping, in which 5 units make up one 5-group and two of the latter a 10-group, or one unit in the next higher column. After the calculation has been 'purified', as this grouping was called in the Middle Ages (the Romans called it *purgatio rationis*), the reckoner can then read off the final number.[1]

The Greeks called the counting board *abax* or *abakion* and the counters *psephoi* (pebbles). Since the word *psephizein* (literally 'to pebble') is also the general word for 'to compute', the Greek language itself provides evidence for the widespread use of the counting board. As late as the fourteenth century the Byzantine scholar Maximus Planudes entitled his textbook on computations with the Indian numerals *Psephophoria kat'Indous* ('Pebble placing in the Indian Fashion'). This suggests that the literal meaning of the ancient expression *psephizein* was still in his mind.

Besides the Salamis Tablet and the passage quoted from Polybius, there is also a third testimony available: a picture of a reckoner at work which appears on the famous Darius Vase. This magnificent ceremonial *crater* (a mixing bowl for wine) in the red-figured Attic style was found in an ancient grave in Apulia and is now in the National Museum in Naples. It was made in the fourth or third century BC. In the central band, the painting represents the council of war of King Darius, before his famous expedition against the Greeks in 490 BC. The person who stands on a platform addressing the king seems to warn him against the dangers of the expedition. A bodyguard stands behind the king. The king's treasurer is seen at his counting board (see Figure 94). To his left and right, there are violently agitated persons who represent the overtaxed tributary provinces.

The Persian treasurer sits at a small table on which are the signs MΨ (Boeotian for X = 1000), HΔΓ and 0 (Boeotian for obol), ⟨ (or ⊂, the sign for the half-obol) and Τ, which has the same meaning as Τ on the Salamis Tablet. A striking feature is the lack of the five-group, for the symbol Γ here clearly stands for the units. Thus, apart from the fractions, only the decimal system is used. In contrast to the Salamis Tablet, the counters or pebbles here lie, without inscribed columns, directly beneath the symbols for the numerical ranks.

FIGURE 94   An artist's representation (enlarged) of the King's treasurer on the Darius Vase (from Menninger, 1970)

In his left hand the treasurer holds a diptych or double tablet, on which is written the word *talanta* H , '100 talents'. On this tablet are posted the amounts which he successively finds on the abacus. Thus a representation of computation in ancient times has been preserved through sheer good luck.

There is a surprising similarity between the Darius Vase and a tiny but exquisitely carved cameo of Etruscan origin. On this gem, in almost exactly the same position and attitude as the treasurer on the Darius Vase, a man is seated at a small, three-legged table on which the counters are represented by three minute spheres. Like the reckoner on the vase, he holds a tablet in his left hand, this time with Etruscan numerals, which leave no doubt as to the man's activity.

The similarity between the representations on two works of art of such divergent origins, which were made at different times, bears witness to the universality of the reckoner. Even his names in different languages, Greek *trapezites* and Latin *mensarius*, 'tabl-er' (from the Latin *mensa*, 'table') show unequivocally that computations were made on tables, just as the present 'bank-er' ultimately derived his name from the medieval 'counting bank' which was his tool.

No specimens of Roman counting boards of the Greek type have survived; but their former existence is clearly indicated by linguistic evidence. The Latin language shows that the Romans used pebbles on a counting board to perform calculations. What the Greeks called *psephoi* the Romans called *calculi*. The Latin word *calx* means 'pebble' or 'gravel stone'. *Calculi* are thus small pebbles. The word *calculare* is of later origin; the Romans themselves used the expression *calculos ponere*, 'placing pebbles', for calculating. When a Roman meant to 'settle accounts with someone' he would say *vocare aliquem ad calculos*, 'to call someone to the pebbles'.

In Wels in Austria, in the old Roman city *Colonia Aurelia Ovilana*, small discs of ivory have been found which probably served as calculi. This is additional evidence for the use of loose, unattached counters.

A different type of abacus, the hand abacus, was used by the Romans for small calculations. Two specimens of the Roman hand abacus are preserved, one in the Cabinet des Médailles in Paris (see Figure 95), the other in the Museo delle Terme in Rome. The figure shows eight long grooves (*alveoli*) and the same number of short grooves at the top. At the right end there is another long groove without the short groove above. On the bridge between the two rows of grooves are the Roman symbols for powers of ten, from ⊠ = one million to | and the symbol O for the *uncia*, and along the extreme right groove the symbols for the half, quarter and third uncia (*semiuncia, sicilicus* and *duella* in Latin). In these grooves run little spheres called *claviculi*, 'little nails', four in each long groove and one in each short groove above.

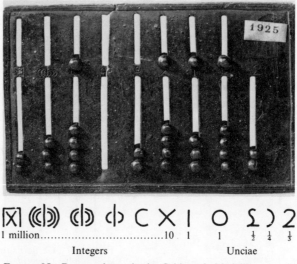

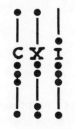

Figure 96   The number 328

FIGURE 95   Roman abacus in the Cabinet de Médailles (Bibliothèque Nationale, Paris). Between the two rows of grooves are Roman number symbols (as illustrated beneath)

Computations on the hand abacus dealt only with the simplest problems, like addition. To insert a number, the counters were pushed towards the raised bridge near the centre. The counters above the bridge are worth five times those below; 8 could be represented as 5 + 3. The number 328 appears in Figure 96.

If a second number was added to the first, the answer was obtained by combining the two. The second number, which the user of the abacus retained in his mind, had to be blended step by step with the first. The numbers 328 and 134 may be added using the hand abacus in the following way. First the spheres are inserted to represent 328, as in Figure 96. Then 100 is added: result 428; next 30 is added, giving 458. Finally 4 is added, to give the sum of 462.

Modified forms of the Roman hand abacus are still in use in Russia, Japan, China and Indonesia, and they will be discussed later. We now turn to medieval forms of the large counting board.

The monastic abacus

It is hardly surprising that information about the abacus declines sharply during the centuries of the decline of Rome and Western Mediterranean culture. Some knowledge of mathematics was preserved and handed on by the late Roman scholars Boethius (died 524) and Cassiodorus (died 570), and by early medieval monks, notably by the Venerable Bede (died 735) and Alcuin (died 804), a teacher in Charlemagne's palace school and the founder of a famous monastery school at Tours. But, until the end of the first millennium, there is no information about computations on the abacus. The abacus may have been in occasional use in certain monasteries, but it was not until around the year one thousand, in the writings of Gerbert, that positive evidence of its use in the Middle Ages first appears.

Gerbert (see also Chapter 4) was born in AD 940 in the Auvergne. He became teacher of mathematics at the cathedral school in Reims. From the evidence of his pupils we know that he took great pains with his teaching and that as a preparation for mathematics he taught computations on the abacus. Later he went to Ravenna and was appointed bishop. In 999 he became Pope Sylvester II. The career of this learned man and his elevation to the Papacy inevitably promoted the spread of his teachings.

Medieval manuscripts describe exactly what Gerbert's counting board looked like. Gerbert himself produced only the 'Rules for Computation with Numbers on the Abacus'. His monastic abacus had parallel columns, which were sometimes closed off at the top by an 'arch' (see Figure 97). This was called *arcus Pythagorei*,

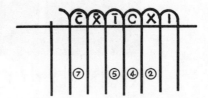

FIGURE 97   Gerbert's reckoning with numbered *calculi* showing the number 705 420

the 'arch of Pythagoras', because in the Middle Ages this Greek was erroneously believed to be the inventor of the abacus. The subscripts of the columns represented powers of ten, as follows:

$$I = 1 \qquad \overline{M} \text{ or } \overline{I} = 100$$
$$X = 10 \qquad \overline{XM} \text{ or } \overline{X} = 10\,000$$
$$C = 100 \qquad \overline{CM} \text{ or } \overline{C} = 100\,000, \text{ etc.}$$

In contrast to the Greeks and Romans, Gerbert did not use pebbles to represent units, nor did he decompose 7 into 5 + 2, as the Romans did on their hand abacus. To represent the number 7 he used a single counter marked with a special symbol meaning 7. As a pupil of Gerbert said 'he used nine symbols, by which he was able to express every number'.

Gerbert's symbols for 1 to 9 were highly peculiar and alien *characteres*, unknown in Christian Europe in the early middle ages. But Hindu-Arabic numerals are recognisable in his characters and in this way they made their first appearance in

Western Europe. Their forms, to be sure, were somewhat strange, at least as they appear in medieval manuscripts.

Gerbert's counters marked with these strange numerals were called *apices*, from the plural of the Latin word *apex*, which means 'tip of a cone'. It is possible that the apices originally had a conical form. Gerbert had some thousand such counters carved out of horn. This was his innovation: the replacement of seven calculi by a single counter or apex marked with the symbol for 7. Where did he obtain these signs? Since it is known that Gerbert spent some time in Spain, where the western Arabs used the so-called gobar figures for their computations, it is more than likely that he became acquainted with them there.

On Gerbert's counting board an apex marked 'zero' was not needed. If there were no multiples of 10 000 Gerbert just left the column $\overline{X}$ empty. The adoption of the Hindu-Arabic numerals without the zero and their use on the counting board was actually a step in the wrong direction. The Hindu-Arabic numerals were destined to *replace* the counting board, making it superfluous; they would not later be used on the counting board.

The apices are less convenient on the board than the pebbles. For example suppose there are 9 pebbles in a certain column on the counting board, and 5 pebbles have to be added. 1 pebble is first added and the 10 replaced by 1 pebble in the next column to the left. The addition is finished by adding 4 more pebbles.

How would the same operation be performed with apices? Suppose the number 29 is displayed on the counting board by means of apices ② and ⑨. 5 has to be added. $9 + 5 = 14$ is performed mentally, then one must take away the apex ⑨ and replace it by ④ and, finally, take away the apex ② and replace it by ③. It is necessary to know the signification of each particular apex, to perform the addition $9 + 5 = 14$ mentally, and to look for the required apices ④ and ③ in a reservoir of apices. The mechanical calculation with pebbles is easier and quicker.

Yet it would be wrong to see in the apices nothing but a useless innovation by Gerbert. The apices made people aware of the existence of the Hindu-Arabic numerals, which were more concise than the traditional Roman numerals.

The apices led a dreary life in the monastic cells of early medieval Europe, and disappeared as soon as the true Indian numerals – this time, properly understood and appreciated – once more found their way to the West.

However, the counting board did not disappear. On the contrary, as production and trade intensified and strengthened life outside the cloisters, the counting board flourished again in a new form. But before going on to discuss this, it is interesting to examine a peculiar method of division on the counting board with *apices*, whose procedure was described by Gerbert.

It is not hard to divide 7825 by 43 in Indian numerals, but the problem becomes very difficult on the monastic abacus. The following method for division was called 'Iron Division'. It was so named because, as a medieval manuscript put it, this method was 'so extraordinarily difficult that its hardness surpasses that of iron'.

To carry out the division of 7825 by 43, the divisor 43 was first increased by 7 to the next multiple of ten, i.e. $43 + 7 = 50$. Then 7825 was divided by 50. The first digit of the quotient is 1, so the first partial quotient is 100. This was multiplied by 50 and the product was subtracted from the dividend 7825, giving 2825. Finally, the error of the first step was compensated by adding 7 times 100, i.e. 700, giving 3525.

This part of the calculation may be written in Hindu-Arabic numerals thus:

$$7825:50 = 100$$
$$\text{subtract } 50 \times 100 = \underline{5000}$$
$$2825$$
$$\text{add} \qquad 7 \times 100 = \underline{700}$$
$$3525$$

Now the procedure is repeated:

$$3525:50 = 70$$
$$\text{subtract } 50 \times 70 \;= \underline{3500}$$
$$25$$
$$\text{add} \qquad 7 \times 70 \;= \underline{490}$$
$$515:50 = 10$$
$$\text{subtract } 50 \times 10 \;= \underline{500}$$
$$15$$
$$\text{add} \qquad 7 \times 10 \;= \underline{70}$$
$$85:50 = 1$$
$$\text{subtract } 50 \times 1 \;= \underline{50}$$
$$35$$
$$\text{add} \qquad 7 \times 1 \;= \underline{7}$$
$$42 \qquad\qquad 181$$

Thus the quotient is 181, and the remainder is 42.

On the counting board the divisor 43 would first be displayed, then the supplement 7, the 'rank' $43 + 7 = 50$ and the dividend 7825. The rank 50 is represented by one apex ⑤ in the column of tens, and the dividend by four apices ⑦ ⑧ ② ⑤. Next 50×100 would be computed mentally and subtracted, replacing the apex ⑦ by ②, and so on.

By performing the calculation on a counting board with four columns, using either pebbles or small discs bearing numbers, one can see why the dividend, and the successive remainders, were divided by 50 and not by 40 as expected. If one divides by 50 the quotient is underestimated, hence the subtraction is always possible. If one divides by 40 the quotient is sometimes overestimated. For instance, 85 divided by 40 would yield 2, which is too much: it results in a negative remainder.

Calculation on the lines

In the late Middle Ages the early medieval monastic abacus with its parallel vertical columns disappeared completely. A look at the illustration (see Figure 98) shows that by 1531 the old counting board with vertical columns had been rotated through a quarter of a turn into one with horizontal lines or strokes. Just when this happened, we do not know: perhaps in the thirteenth century. The board with horizontal lines had definitely been long in use by the eighteenth century for all books and pictures of that date show the columns horizontal and contain not a word about the vertical monastic abacus.

FIGURE 98   Sixteenth-century
woodcut by Hans Weiditz,
Augsburg, 1531, showing a clerk
with his reckoning table and his
book (from Menninger, 1970)

Why this change in orientation? Probably because it was more comfortable to 'read' the counters horizontally. Long horizontal rows are easier to grasp than long vertical columns. Because of the restoration and great expansion of trade about the twelfth century (as a result of the Crusades and the Hanseatic League, among other things), there was an enormously increased demand for money changing. Everyday computation, which had been reduced to a small trickle at the time of the monastic abacus, now began to flow again in a mighty stream. Merchants, shopkeepers, and officials may well have adopted the monastic abacus or been guided by it; at any rate, they improved it and made it handier, among other things by turning it to the horizontal position.

But merchants and money changers again took up the old undifferentiated calculi, and thus did not adopt Gerbert's innovation of apices with different values. They would represent the number 4 no longer with one counter bearing the numeral 4 but again by putting down 4 counters of equal value. In addition to counters made from the usual materials such as bone, wood, and metal, from the thirteenth century on there were stamped or embossed counters which, however, had no monetary value. Some of these counters bear heraldic arms, or portraits; others even depict reckoning tables.

The horizontal reckoning table existed in two forms. The first, the so-called 'line board' had lines without any indication of value, upon and between which the counters were placed; the other had specified 'coin rows', each for a specific monetary unit, within which the counters were laid out. The latter form was used for calculations involving sums of money and the conversion of one monetary unit into another. The line board, on the other hand, was intended for computations with abstract, unnamed numbers, and could also be used for multiplication and division.

Inventories and wills show that in the later Middle Ages reckoning boards were used in a great variety of places: in monasteries, in royal treasuries, in the offices of town officials and in the counting rooms of merchants. Only a few of these reckoning tables have been preserved; but thousands of counters have survived. The reason is clear. The counters, being intrinsically useless and made of base metal, outlived the period in which they were used and at the same time escaped the melting pot. But the tables were of wood. Once they were worn out as computation instruments, they could still serve as ordinary tables, until they finally ended up by being burnt.

The earliest counters, dating from the thirteenth century, come from France. The latest German counter comes from the Board of Mines of the Harz Mountains, which used counting boards for their business until the beginning of the eighteenth century. Between these limits of time the new reckoning board was extremely popular.

The counting board and the counters are often mentioned in literature. For example, there is the fine passage in Shakespeare's *The Winter's Tale* (Act IV, Scene 2), where the young shepherd has to work out a sum:

Let me see: every 'leven wether tods; every tod yields pound and odd shilling; fifteen hundred shorn, what comes the wool to?

and then cries in despair:

I cannot do't without counters.

With the same purpose in mind, Luther mentions them in one of his colourful sayings:

Thus the Jews placed the counters on the lines and reckoned how many Canaanites there were and what a small number of Israelites there were.

From the sixteenth century on there were a lot of books in existence which taught the reckoning on the lines. A very popular and widely read book in England was *The Grounde of Artes*[2] of 1541, in which Robert Recorde dealt among other topics with arithmetic computations. Another well-known English book was the St Albans Book of Computations of 1537: *Introduction for to Lerne to Reckon with the Pen or with the Counters.*

As a last piece of indisputable evidence against the view that computations in the old days were made with Roman numerals, the Bavarian reckoning cloths may be mentioned. These cloths had the same division as the reckoning tables, but they could easily be carried about by the officials whose duty it was to check the calculations made by the mayors of the towns and other administrative centres throughout the province. Along with these counting cloths, there is a contemporary document which very clearly reveals their use:

Notice   This cloth was formerly used by the provincial inspectors of finances in their annual inspection of computations in which everything was calculated in black counters with gulden, pounds, shillings, and pennies, as follows: the Lord Mayor would read the amounts and call out the numbers, for example, 10·gulden, 5 shillings and 2 pennies. The first or the second Cavalier had the cloth and a dish of silver counters before him. As the mayor called out the amounts, the Cavalier would place a silver counter in the space marked 10 gulden, 5 counters in the space for the shillings and 2 counters in the space for the pennies. Then they would proceed in this manner. But after there were 10 counters in the 10-gulden space, he would take these away again and put down one in the space where the hundreds were marked. But if there were 7 counters in the shilling space, these were also removed and converted with one counter into 1 gulden, which could always be multiplied by the number ten, hundred, or thousand. The same was done with the pennies, so that as often as there were 30 counters together in the pennies' space a shilling was made, and from the latter likewise a gulden. The second Cavalier had the certificates or proofs at hand for his examination. The prelate, however read the second item or, if it was a simple matter as from the Lowland, the finished computation. As soon as the space in which a sum was placed was complete, the Cavalier would so state, according to the information on the reckoning cloth and the counters lying on its spaces. But it was the mayor's duty to see that this was consistent with the written accounts.

The actual preparation of the line board was as follows: four parallel horizontal lines were drawn on a board or table, with a vertical line down the middle dividing them into two columns or *bankire* (see Figure 99). The topmost of the four

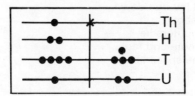

FIGURE 99   Basic form of the line board

horizontal lines was marked with a cross at the centre. Unlike the reckoning board with spaces for denominations of coins, the counters were here placed directly upon the lines themselves. These were unspecified, for they imparted to the counters only an abstract decimal place value, independent of any specified system of coinage or weights: a counter on the bottom line had the value of 1, on the second line 10, on the third line 100, and on the top line, which was usually marked with a cross, 1000. The arithmetician Johann Albrecht of Wittenberg had this to say about this technique in his *Rechenbüchlein auff der linien* ('Booklet of computations on the Lines') of 1534:

> So that the lines may be known / they are to be marked in the following manner: the line that is called the first means one / the next line about it ten / the third hundred / and the fourth thousand / mark this last line with a small cross / and count off on the same line again (as on the first) one / on the second above ten / on the third hundred / and on the fourth thousand. But mark this one with a small cross. But starting with the first small cross / for each line you must say thousand. When there are one thousand / ten thousand / a hundred thousand / a thousand times a thousand / and as many small crosses as there are, so many thousands must you always say. You must also know / that any space means five times as many as the line next beneath it (to which it belongs), and the space under the first line / means a half...

In the left column of the line board shown the number 1241 is displayed. In the right column a pebble is placed in the space between two lines. This pebble replaces five pebbles on the line just below it. Thus the number placed on the right is 82. A counter placed beneath the bottom line means $\frac{1}{2}$. In this very simple manner the more elegant representation of numbers with the aid of the quinary 5-grouping was reintroduced after its abandonment by the early medieval abacus with its numbered apices.

Many textbooks of the time list the following arithmetical operations: *numeration* (the placing of the numbers on the counting board), as in Figure 100, *addition*, *subtraction, duplation* and *mediation* (that is, the doubling and halving of numbers), *multiplication*, and *division*. Along with each of these operations, there was *elevation* or *purification*, the grouping of units into smaller numbers of higher units. Together with subtraction there was *resolution*, the 'de-grouping' of higher units into lower ones. *Elevation* and *resolution* together constituted the operation of *reduction*. These operations may be explained as follows (see Figures 100 and 101):

1. *Addition* What is the sum of 3507 and 7249?

(a) *Numeration* Both of these numbers are placed with counters on the board.

(b) *Elevation* Explained in a German medieval text as follows:

> See where two counters lie in the space
> Pick them up and lay one down
> On the next line above
> Likewise if five counters lie
> On one line take note and look
> And put one in the space above.

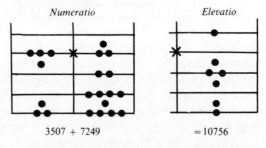

FIGURE 100 The sum 3507 + 7249 = 10756 as worked out on the line board

If several numbers are to be added, one begins by adding the first two, then one adds the third, and so on.

2. *Subtraction* A subtraction on the line board consists of three steps:

(a) *Numeration* Place the given numbers on the fields

(b) *Resolution* Replace some of the pebbles by 2 or 5 pebbles on a lower level, in order to make the subtraction possible.

(c) *Subtraction* Draw away as many counters from each row of the first field as there are counters in the corresponding row of the second field.

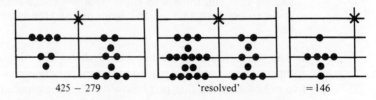

FIGURE 101 Subtraction of 425 − 279 = 146 on the line board

The 'resolution' in the example in Figure 101 can be explained as follows. Starting from below, the single pebble between the first and second line is replaced by five pebbles on the first line. Next, one of the two pebbles in the second line is 'resolved' into two pebbles between the first and second lines. Next, one of the four upper pebbles is resolved into two lower pebbles between the lines, and one of the two is again replaced by five pebbles on the second line. Now the subtraction can be performed.

3.   *Doubling and halving*   These operations were introduced as separate independent procedures by the Master General of the Dominican Order, Jordanus Nemorarius (died *c*. 1236). After his time they frequently reappeared in the writings of arithmeticians.

The counting board directly promotes doubling and halving, because these procedures can be carried out without any computation at all, simply by adding or removing counters according to a fixed rule.

The rule for doubling is obvious: one just doubles the number of pebbles on every line and also between the lines. The reverse procedure is halving. The German textbook writer Adam Riese (sixteenth century) described it as follows:

> Always for two counters place one over into the other field and likewise on the line on which you have your finger place one counter over into the other field. And for one counter alone you shall place one in the next space beneath the line on which you have your finger. But where there is no counter none is transferred. And that is all there is to halving.

Adam Riese did not say how the single pebbles between the lines were to be treated. However, if each of these single pebbles is replaced right at the beginning by five pebbles on the line just below, no problem arises.

Doubling and halving are primitive operations, being early forms of multiplication and division. They occur even in ancient Egyptian texts, as shown in Chapter 5.

4.   *Multiplication*   This was taught by Adam Riese in three steps:
(a)   The multiplier has one digit, for instance 28 × 6. Adam Riese wrote:

> Know that you have two numbers to be multiplied. One, which is to be multiplied, lies always on the line [on the left]. The other number with which you wish to multiply write down.
> Now if you are to multiply with one digit [e.g. with 6] then take the highest line on which one or more counter lies and place your written number [in our case 6] as many times as there are counters lying on the same line. But when a counter lies within a space [between two lines], take the next line above the same space and put down only half the number you have written down.

(b)   The multiplier has two digits, for example 28 × 34:

> But if you wish to multiply a number by one which has two digits, take the next higher line above the counters and place the other digit of the number you have written down [in our case 3, the 'other digit' of 34] as many times as there are counters on the line below. [This means: multiply the given number by 30.] Next take the line on which the counters lie, and put down the 'first' digit of the written number [in our case 4] as many times as there are counters lying upon the line. [This means multiply the given number by 4. The product, of course, is added automatically to the earlier result.]

(c)   The multiplier has more than two digits.

> So proceed also with three, four, five and more digits, always placing the fifth digit of your written number on the fifth line from the line on which your counters lie, and begin counting by placing your finger on the line. Place the fourth number on the fourth line, and so on to the bottom line. But with [pebbles lying within] the spaces proceed as shown in doubling.

Then follows an important admonition:

> Take care to learn your multiplication table well, and you will master all computations.

This means learning the multiplication table from 1×1 to 4×9 by heart.

5.   *Division*   The division by a number having one digit only is explained by Adam Riese as follows:

> Place the finger of your left hand on the top line [of the dividend], and see whether you can take away the number by which you wish to divide. If you cannot take it away, place the finger on the next line, and do this as long as you can take away the number by which you wish to divide. Take it away as often as you can, and each time place a counter at the finger. Do this until you can no longer take away the number.

This is how 44 would be divided by 3. First display 44. From the four counters on the upper line take away 3 counters. Place one counter next to your finger on the upper line. Now place your finger on the lower line and remove four times 3 counters, every time placing one counter next to your finger. Quotient 14, remainder 2. This example therefore offers at least a glimpse of ordinary computations 'on the lines', as they were performed by townsmen and traders in the late Middle Ages.

\ Köbel and other arithmeticians also taught more difficult computations on the line board, such as currency conversions and the extraction of square roots and cube roots.

The Chinese, Japanese and Russian abaci

The history of written numbers contains one instance in which a system of numerals derived from the counting board achieved universal validity and recognition. Since the era before the birth of Christ the Chinese used little bamboo or wooden sticks as calculating pieces (*chou*) on their counting board (see Chapter 4). To form the numbers the Chinese arranged them in a special manner (see Figure 102). Firstly, the arrangement of the digits was changed in successive columns: the units, hundreds and ten thousands were placed vertically and the tens and thousands horizontally. Secondly, they used a quinary grouping for greater ease in reading the 'digits'.

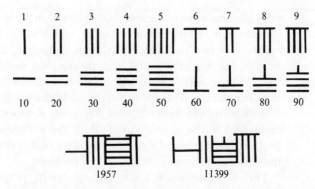

FIGURE 102 *Sangi* numerals made up from sticks; from the tenth century AD the figures were condensed into monograms as shown in the bottom line

There is a reference to these stick numerals in the following note in the *Sun Tzu Suan Ching* (*c.* AD 300):

In making calculations we must first know the positions and structure [of numerals]. The units are vertical and the tens horizontal, the hundreds stand while the thousands lie down; thousands and tens therefore look the same, as also the ten thousands and the hundreds.... When we come to 6 we no longer pile up [strokes], and the 5 has not got one.[3]

The demarcation of neighbouring powers of ten in this way enabled the computers to use counting boards without marked vertical columns. Before the eighth century AD the place where a zero was required was always left vacant. Afterwards in written manuscripts a circle was placed to note an empty space in a column.

There is a lot of evidence that from the Warring States period (fourth and third centuries BC) calculations were made with stick numerals on a counting board, using a place-value system. Stick numerals are depicted on the coins of that period; they are also described in literary works. Perhaps the most famous instance of the latter is in the *Tao Tê Ching*, where Lao Tzu says 'Good mathematicians do not use counting-rods'.[4] They were not only employed for ordinary computations but even for solving algebraic equations.

Following Needham's exposition, one can see how the multiplication 81 × 81 was performed on the counting board (see Figure 103). Needham has chosen just this

FIGURE 103 The multiplication 81 × 81

example, because it is explained in detail in Chapter 1 of the *Sun Tzu Suan Ching* (*Master Sun's Arithmetical Manual*), which was written in the late third century AD.

First, the two factors were laid out in the top and bottom lines. Then, 'the upper 8 calls the lower 8', and the product 64 is laid out two spaces to the left, representing 6400. Next, the upper 8 'calls' the lower 1, and 8 is laid out in the middle position. The upper 8 is then withdrawn: it is no longer needed. The upper 1 is 'called' by the lower 8, and the lower 8 withdrawn. Finally, '1 calling 1' is set down as 1 in the units column. With the addition of the partial products the final product is obtained. Of course, the user of this procedure had to know the table of multiplication from 1×1 to 9×9 by heart.

The Chinese used the counting board not only for addition, subtraction, multiplication and division of numbers, but even for the numerical solution of algebraic equations such as

$$x^3 = 1\,860\,867$$

or

$$x^3 + 34x = 71\,000$$

The Chinese method for solving such equations is explained in the *Nine Chapters on the Mathematical Art*, which were probably written in the first century BC. The Chinese method is essentially the same as that of Horner. The method was rediscovered by Ruffini in 1802 and by Horner in 1819.

Around AD 600 the Japanese adopted the counting sticks and used them until quite recently. Of particular interest here, however, is the *soroban* of Japan (see Figures 104 and 105), which is an exact analogue of the Roman hand abacus. It consists of a rectangular frame housing thin rods (Japanese *keta*, 'row', 'wire'). Six small beads slid along each rod, the sixth being separated from the other five at the bottom by a long strip perpendicular to the vertical wires or rods. The differences between the Japanese *soroban* and the Roman hand abacus are that instead of four counters at the bottom there are generally five; the vertical rods or wires have no numerical symbols which are unspecified; the *soroban* has more vertical columns, as many as seventeen; and it has no columns especially set aside for fractions. They do have in common, however, the quinary and decimal grouping of the counters which run along fixed rods (or grooves). On the *soroban*, as on the Roman hand abacus, only the counters pushed against the bridge or separating bar have value. In Figure 104 it can be seen that the numbers 231 and 1956 are formed on the *soroban*.

FIGURE 104   The Japanese *soroban*; the number 231 is shown in the middle, and 1956 on the right (from Menninger, 1970)

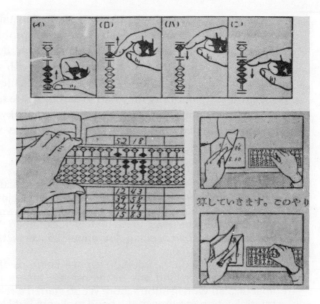

FIGURE 105   Stages of using a *soroban*. Top, showing the proper
technique for moving the counters along the *keta*; bottom left,
placing the number 5218 on the *soroban*; bottom right, how the
merchant adds the posted amounts together (from Menninger,
1970)

The *soroban* was introduced into Japan from China, probably in the sixteenth
century. The origin of its name is unexplained. In the 1870s it was almost replaced
by written computations with the 'western' (i.e. Indian) numerals, which had by
then come into universal use throughout the world, but since about 1930 it has
again been in great demand as a result of the growth of Japanese industry and
trade. The Japanese chambers of commerce and industry offer yearly examinations
and competitions in computations on the *soroban* in which, to name only one year,
in 1942 about 40 000 participants competed for the prize.

In modern Japanese *soroban* textbooks all four operations of arithmetic are
taught (see Figure 105), especially multiplication and division. The many *ketas* or
wires permit all the numbers in a computation to be recorded, as for example, both
the factors 27 × 6 and also the result 162 (see Figure 106). As always on an abacus,
the rules of place value are extremely important. This is made clear by an example:
the factor 6 is placed on the left, and 27 under ketas A and B. The multiplication 6
× 7 = 42 is performed and placed on keta D, and the number 7 is thereby
removed from keta B. Then follows 6 × 2(0) = 12(0), which is immediately
combined with 4 into 16; final answer: 162

To carry out this procedure, the reckoner must know the multiplication table, the

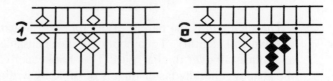

FIGURE 106   27 × 6 = 162 on the *soroban*

rules of place value and the proper method of instantaneously combining numbers. In other words, he must constantly work out problems in his head.

Great importance is attached to elementary education in Japan, including the emphasis, surprising to us, on computations with the *soroban*. When the Americans occupied Japan in 1945, they at first ridiculed the *soroban* for its supposed backwardness. Naturally they had to demonstrate their own 'progressive' methods, and so they organised a calculating match in Tokyo, which was observed by 3000 spectators. This competition, which was in many respects one between two very different cultures, had a surprising outcome.

> The contest matched 22-year-old Kiyoshi Matsuzaki, a Japanese Communications Ministry clerk with seven years' special abacus lessons, against 22-year-old Thomas Ian Wood of Deering, an Army finance clerk with four years' experience on modern machines. Matsuzaki, who flipped the wooden beads with such lightning dexterity that he was immediately nicknamed 'The Hands', used an ordinary Japanese soroban, selling for about 25 cents before the war. Wood's electric machine cost $700.
>
> The abacus won the addition event – columns of four-to-six-digit figures – taking all six heats finishing one of them more than a full minute ahead of Wood. The abacus also won in subtraction. Wood staged a rally in multiplication, since abacus multiplication requires many hand motions; but Matsuzaki was out in front again in division, and in the final composite problem. 'The Hands' also made fewer mistakes.
>
> One reason why Matsuzaki won is that like all abacus veterans, he does the simplest arithmetic in his head, pegging the results on the abacus and going on from there.[5]

The Chinese *suan pan* (literally, 'reckoning board'; see Figure 107) has two beads on the separated 5-group wires instead of the single one on the Japanese *soroban* and the Roman hand abacus. This improvement makes addition a little easier: 6 + 8 = 14 can be first formed on one wire or rod and then added together, whereas on the *soroban* they must be combined immediately. Like slide rules, some are almost a foot and a half long, while others may be no larger than a matchbox. As in Japan, children in China are instructed in the use of the *suan pan*.

FIGURE 107   Chinese *suan pan* (Science Museum, London)

The *suan pan* is the parent of the *soroban*, which was introduced into Japan in the sixteenth century. Today both are indispensable aids to computation in shops and offices all over the Far East. The shopkeeper can add up several numbers on the abacus generally faster than is possible by writing them down. And just as an experienced clerk in a Western bank depresses the keys of an adding machine without looking at them, the Chinese and the Japanese play on their *suan pan* and *soroban*, pushing the counters back and forth with a light 'click-click' instead of the rattling noise made by the machine.

The *suan pan* did not appear in China before the twelfth century. Whether it was an independent invention or an adaptation of the Roman hand abacus is not known.

There is still another form of the abacus with perpendicular wires, on each of which ten beads are strung. On these the fifth and sixth beads are coloured differently from the rest, to make counting easier. These are the Russian *ščëty* (Russian *ščët*, 'counting, reckoning', nominative plural *ščëty*). A Russian estate owner, always had his *ščët* at hand on his desk. He used it for computing anything that could possibly be computed.

Figure 108 shows a *ščët* from Persia, where it was used (as in Turkey) by merchants and tradespeople. It is still used by older people. Of all the forms of the abacus, this one has been the most 'popular', the one most readily accepted. This is because it is easy to use in theory, and requires no previous schooling or long practice as does the *soroban*. Also, anybody needing one can easily make one. However, its capacity is more limited than the *soroban*.

FIGURE 108   Russian *Ščët* from
Persia (Science Museum, London)

Anyone who calculates on the *ščët* soon comes to recognise the value of specially marking out the two middle beads in the row of ten. There is no doubt that this small calculating device is one of those simple but highly significant inventions which, like the wheel, have enabled humankind to overcome hitherto impossible obstacles.

Perhaps when the reader first sets eyes on a picture of a *ščët*, he or she is struck by the thought: 'Why, this is just the small abacus our baby plays with'. Surprisingly, this is absolutely right: a child's abacus is a descendant of the Russian one. During his invasion of Russia in 1812, Napoleon had in his army an engineer with the rank of a lieutenant, Poncelet, who later became famous as a geometer. He was one of the founders of projective geometry. During the French retreat, Poncelet was captured by the Russians and brought to Saratov on the Volga. While there, living among simple people, he became so impressed with the excellence of the *ščët* as a device for teaching children that, upon returning to France, he introduced it into all the schools of the city of Metz. From here the *boullier* (from the French word *boule*, 'ball') spread all over France, Holland and Germany and even America. Thus it may well be that the Roman hand abacus has come down to us by way of a long detour through Asia.

Figure 109 shows the development of the counting board from classical antiquity to modern times. The number 2074 is shown throughout.

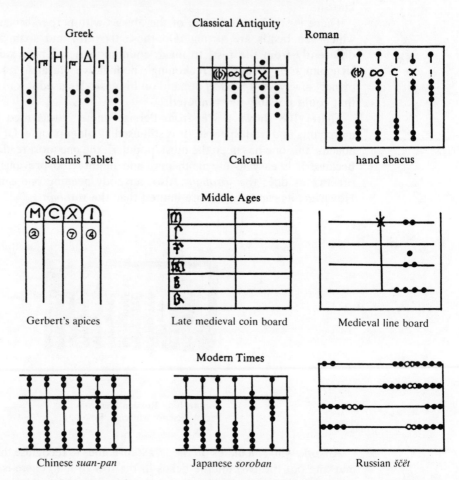

FIGURE 109   The development of the counting board from classical antiquity to modern times

Early calculating machines

By the beginning of the seventeenth century the victory of the Hindu–Arabic system of numeration for both calculation and recording was complete in most of Europe, and the abacus went out of use west of Russia. Even so, it was a long time before even the basic processes of calculation became commonly understood or widely practised. Thus, for example, Samuel Pepys – a senior civil servant in the Admiralty and one of the best educated and most intelligent men of his time – recorded in his diary (4 July 1662) how he rose at 5 a.m. to learn 'mathematiques' from a Mr Cooper, 'he being a very able man'. Pepys continues:

> After an hour's being with him at arithmetique (my first attempt being to learn the multiplication table); then we parted till tomorrow.

Difficulties associated with multiplication – and still more, division – were resolved by the invention of *logarithms* by John Napier of Merchiston, near Edinburgh. The book in which he described his system and gave the first tables of logarithms was published in 1614 with the title *Mirifici logarithmorum canonis descriptio* ('A Description of the Marvellous Rule of Logarithms'). The effect was to reduce all arithmetical calculations to addition and subtraction. The importance of the new technique was soon recognised by the practising mathematicians of the day and within a few years steps were being taken to mechanise the process. The *slide rule* appeared in about 1630. Pepys bought one in 1663 and found it 'very pretty for all questions of arithmetic'. Pepys's approval has certainly been endorsed by posterity; the slide rule was the 'badge of office' of the engineer for the better part of the last hundred years, though it has now been largely replaced by the electronic calculator.

What may be called the 'mechanical era' has now been reached. The mechanical desk calculator originated in the seventeenth century; the two crucial inventions were both made during Pepys' lifetime by those towering geniuses, Pascal (1623–62) and Leibniz (1646–1716). Pascal completed his first calculating machine when he was nineteen years of age; he built it, so the story goes, to assist his father, who was a tax collector. He 'advertised' his invention thus.

> I submit to the public a small machine of my own invention, by means of which you alone may, without any effort perform all the operations of arithmetic, and may be relieved of the work which has often times fatigued your spirit when you have worked with the counters or with the pen.

Pascal's machine (see Figure 110) was an adding and subtracting device; multiplication had to be achieved by repeated addition. Numbers were set by a stylus on one group of wheels and the results read from another set of wheels through windows on the cover plate. The machine contained several ingenious features, including a ratchet device for transmitting the 'carry' from one digit-place to the next.

Pascal, then, had mechanised the process of addition and subtraction. Leibniz, in about 1671, took the logical next step and mechanised multiplication. He explains how his machine (see Figure 111) consists of two parts, one designed for addition and subtraction – which coincides exactly with the calculating box of Pascal – and the other for multiplication and division: Leibniz then describes how the machine has three kinds of wheels: 'the wheels of addition, the wheels of the multiplicand and the wheels of the multiplier' (see Figure 112). After explaining how they work he mentions that to enable multiplication to be performed quickly and easily a peculiar arrangement would be needed the exposition of which would lead too far into details. This 'peculiar arrangement', known as the *Leibniz stepped wheel*, is a crucial feature of the whole design. It is, in effect, a gear wheel containing a variable number of teeth and it remained in use, essentially in the form Leibniz left it, until the manufacture of mechanical calculators practically ceased a few years

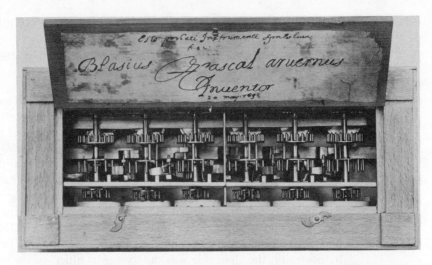

FIGURE 110   Two views of the calculating machine built by Pascal in 1642 (Science Museum, London)

ago. Leibniz ends his account on a buoyant note:

> And now we may give final praise to the machine and say that it will be desirable to all who are engaged in computations which, it is well known, are the managers of financial affairs, the administrators of others' estates, merchants, surveyors, geographers, navigators, astronomers, and those connected with any of the crafts that use mathematics.

The story of the desk calculating machine since the time of Pascal and Leibniz has been one of slow but steady improvement in detailed design and production methods. It was not until about 1810 that the first successful commercial machine was made in quantity by C. Thomas of Colmar, Alsace (see Figure 113). By the beginning of this century, the mechanical desk calculator had become a familiar object in both commercial and technical establishments. Recently they were superseded, in their turn, by the pocket-size electronic calculator.

The automatic computer

The calculating devices so far discussed are *non-automatic* in the sense that they require frequent attention by a human operator – to move the beads of an abacus,

FIGURE 111   Two views of Leibniz's calculating machine (Science Museum, London)

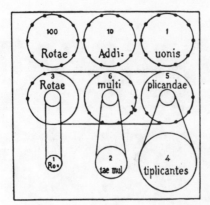

FIGURE 112   The principle of
Leibniz's stepped wheels for
addition, multiplication and
division using in his calculating
machine (from A. Smith, *Source
Book in Mathematics*, Vol. 1,
courtesy of Dover, New York)

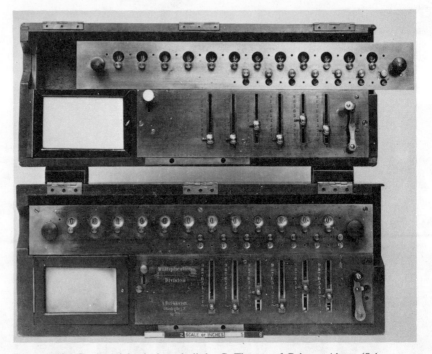

FIGURE 113   Commercial calculator built by C. Thomas of Colmar, Alsace (Science
Museum, London)

to press the keys or crank the handle of a mechanical calculator, to record answers
on paper, and so forth. This method of working clearly imposes a severe limit to the
speed of calculation and prompts the question: Is it possible to make an *automatic*
calculating machine, i.e. one that can carry out extensive calculations without
human intervention?

The first man to put forward detailed proposals for such a machine was Charles
Babbage (1791–1871), one of the founders of the Analytical Society in Cambridge.
He was born in Devon, the son of a wealthy banker, and graduated at Cambridge in
1814. He passed the rest of his life as a cultivated gentleman of independent means
pursuing his many interests, riding his hobby horses, travelling widely, meeting all

the best people and quarrelling with many of them. In 1828 he was appointed to the Lucasian Chair of Mathematics at Cambridge; he resigned in 1839 to devote himself to his mathematical machines. He found time for much else, however: to attempt to reform the Royal Society, Greenwich Observatory and the teaching of mathematics at Cambridge; to analyse the operation and economics of the Post Office, the printing trade and the pin-making industry; to publish one of the first reliable actuarial 'life tables'; to stand for Parliament; and to make some of the earliest dynamometer measurements on the railways, running a special train on Sundays for the purpose. His most successful book, *Economy of Manufacturers and Machinery*, was published in 1832 and was widely translated. He was intensely curious, interested in everything, full of ingenious ideas and practical inventions of all kinds. In short, a most fascinating character in the tradition of the nineteenth-century gentleman-eccentric! His main claim to the attention of posterity un-doubtedly stems from his life-long devotion to his mathematical machines – from his obsessive idea that heavy calculations could be 'executed by steam', as he once put it to a friend.

In 1812 Babbage conceived the idea of constructing what he called a *Difference Engine* (see Figure 114), a mechanical device for computing and printing tables of mathematical functions. (Existing tables were notoriously inaccurate.) The mathematical basis of the device is a procedure known as the *method of differences* which enables a table of values of any polynomial function to be built up, step by step, by simple additions only. Such an engine must consist of a number of registers (Babbage suggested six) linked together so that additions from one register to the next may be performed according to a fixed cycle. Babbage first built a small pilot machine which he demonstrated in 1822 to such effect that he secured the support of the Government and the Royal Society for the construction of a full-sized Engine. The machine was never completed and official support was finally withdrawn in 1842, after the Government had contributed £17 000 and Babbage a substantial part of his private fortune.

Some years later, a Swedish engineer, Georg Scheutz, and his son succeeded in building a working machine based on Babbage's ideas (see Figure 115). A copy of the Scheutz machine was made for the British Government in 1864 and used in the Registrar General's office to prepare life tables. Several types of mechanical Difference Engines have been built commercially since.

Although the Difference Engine project was not finally abandoned until 1842, Babbage had largely lost interest in it ten years earlier, when his imagination had been fired by a much more ambitious project for the construction of what he called an *Analytical Engine*. Although this machine, like the earlier one, was never built, Babbage's ideas are of the greatest interest today because the Analytical Engine was conceived – and also extensively designed and partially built – as a completely automatic *general-purpose* computer. The Difference Engine, by contrast, was a *special-purpose* device. In the Analytical Engine Babbage provided for the five essential constituent parts which we now know any automatic general-purpose computer must have. These, as shown in Figure 123 (p. 210), are:

1. a *store* for holding numbers – both those forming the data of the problem and those generated in the course of the calculation (in modern computers the store holds the operating instructions as well);

2. *an arithmetic (or processing) unit* for performing arithmetic or logical operations on those numbers;
3. *a control unit* to cause the machine to perform the designed operations in the correct sequence (this performs the role played by the human being in non-automatic calculation);
4. *input devices* whereby numbers and operating instructions can be supplied to the machine;
5. *output devices* for displaying the results of a calculation.

FIGURE 114 Babbage's Difference Engine, a replica of the first two machines he built, neither of which was completed (Science Museum, London)

For storage Babbage proposed to use columns of wheels, each wheel capable of resting in one of ten different positions. Transfer of numbers between the store and the arithmetic unit were to be accomplished by means of an elaborate mechanism of gears, rods and linkages. The store itself was to accommodate one thousand

FIGURE 115   Georg Scheutz's difference engine (Science Museum, London)

numbers, each number being represented to fifty decimal places. (A typical piece of Babbage extravaganza!) He had several other equally ambitious ideas, e.g. for feeding in mathematical tables in the form of holes punched on cards, and for automatic printing of the Engine's results.

Babbage's ideas for controlling the operation of the Engine are particularly interesting in the light of how it is done today. He proposed to adopt the method employed to control the Jacquard looms, which were used for weaving fabrics of complicated design. During each throw of the shuttle which carries the weft threads, certain of the warp threads must be lifted, in accordance with the requirements of the pattern, so that the shuttle may pass underneath. In 1801 J. M. Jacquard (1752–1834), a master weaver of Lyons, invented a system of controlling the threads by means of punched cards (see Figure 116). The ingenuity of this invention fired Babbage's enthusiasm. He wrote:

> The Jacquard loom weaves any design which the imagination of man can conceive. The patterns designed by artists are punched by a special machine on pasteboard cards and when these cards are placed within the loom, it will weave the desired pattern.

A portrait of Jacquard, woven on one of his own looms and requiring no fewer than 24 000 cards to produce it, hung in Babbage's drawing room. Most people thought it was an engraving.

Babbage proposed to use two sets of cards in the Analytical Engine: the *operation cards* (to control the action of the arithmetic unit) and the *cards of the variables* (to control the transfer of numbers to and from the store).

The Analytical Engine, like the earlier machine, was, alas, never completed although Babbage continued to work on it throughout his life. The basic trouble was that his schemes were far too ambitious. It is often said that the Engine could not be built because the techniques of precision engineering in the nineteenth century were inadequate to meet his demands. However, we must remember that superb watches and clocks were made at this time. The crucial point is that any device which uses metallic moving parts cannot operate at high speed because of

FIGURE 116   Jacquard loom (courtesy IBM)

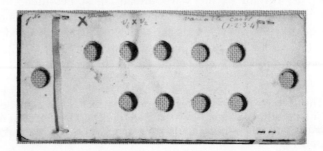

FIGURE 117   Babbage's operation card (Science Museum, London)

the inertia of those parts. The speed of modern computers is possible because they do not have such moving parts – they are electronic.

As the years went by the frustrated inventor became increasingly embittered by a sense of failure. He suffered the unhappy fate of the genius who is too far ahead of his time. It has taken the world a century to catch up with him!

Although Babbage wrote extensively on many topics, he left no systematic account of the Analytical Engine. Fortunately we are well served from other sources. In 1840 Babbage was invited to Turin to attend what he called 'a meeting of Italian Philosophers' and he gave a talk to a group of engineers and mathematicians. Among his audience was L. F. Menabrea, a young engineer officer on the staff of the Military Academy of Turin who later became one of Garibaldi's generals. Menabrea wrote an account of Babbage's ideas and published it in a

Geneva journal in 1842. The paper was translated into English by Lady Lovelace, who added extensive notes of her own (three times as long as the original), and it was published in Taylor's *Scientific Memoirs* in 1843. In his autobiography, published some twenty years later Babbage wrote:

> These memoirs furnish to those who are capable of understanding the reasoning a complete demonstration that the whole of the development and operations of analysis are now capable of being executed by machinery.

Ada Augusta, Countess of Lovelace (1815–52), was the only child of Lord and Lady Byron. She had considerable mathematical talent – perhaps inherited from her mother, who was described by her husband as the 'Princess of Parallelograms' – and frequently visited Babbage while he was working on his engines. She acquired a thorough grasp of Babbage's ideas and was able to explain them far more clearly than Babbage himself was ever able to do. (Her gift of language was presumably inherited from her father!) She invented the idea of a *program loop*, vital to the computer today.

Both Menabrea's paper and Lady Lovelace's notes are mainly concerned with what is now called the subject of *programming* – the formulation of a schedule of instructions (the program) which, when executed, will cause the machine to carry out a desired calculation automatically. Lady Lovelace goes into the subject in considerable detail and illustrates her points by describing several programs for performing quite advanced mathematical calculations.

The next major advance – the realisation that the positions of holes in a Jacquard type card could be detected by electrical means – occurred some twenty years after Babbage's death. *Hermann Hollerith*, a statistician on the staff of the US Bureau of the Census, experimented with devices working on this principle for counting and sorting, and some of his machines (see Figure 118) were used for analysing the US Census of 1890. The range of 'Hollerith' machines was soon extended to deal with most of the operations of office arithmetic and punched card installations – groups

FIGURE 118   The Hollerith horizontal sorting machine (Science Museum, London)

of single-purpose electromechanical devices – were to be found in many large commercial organisations during the first half of the twentieth century.

The fully automatic general-purpose computer was therefore conceived in Cambridge, England, in 1832 and was born in Cambridge, Massachusetts, 112 years later! In 1937, Howard Aiken of Harvard had the idea of using the techniques and components developed for punched card machines to produce a fully automatic general-purpose calculating machine. To this end he approached the International Business Machine (IBM) Corporation – one of the largest manufacturers of punched card equipment. The result of their collaboration was the *Automatic Sequence Controlled Calculator* (*ASCC*), which was completed in 1944 and presented to Harvard College where it remained in continuous use for some fifteen years. Babbage's dream had come true at last. The ASCC was built on a lavish scale with no expense spared (see Figure 119). It contained more than three-quarters of a million parts and use more than five hundred miles of wire. Although a very slow machine by today's standards (0.3 s for addition; 4 s for multiplication), it has a unique claim to fame as the first fully automatic computer to be completed.

FIGURE 119   IBM Automatic Sequence Controlled Calculator at Harvard University, 1944; known as Mark I this was the first electromechanical computer and was designed by Howard Aiken (courtesy IBM)

A further major advance – the application of electronic techniques to computer design – followed very quickly. The first electronic digital computer (see Figure 120), the *Electronic Numerical Integrator and Calculator* (*ENIAC*), was completed in 1946 at the University of Pennsylvania at Philadelphia. In saying that ENIAC was an electronic computer we mean that the storage and manipulation of numbers inside the machine, and also the control of the sequence of operations, were done by means of electronic circuits. Indeed, apart from the input and output mechanisms, the machine had no moving parts; some 18 000 thermionic valves were used instead! The immediate effect of the introduction of electronics was an increase in operating speed of more than one-thousand-fold; two ten-digit numbers could be added in 1/5000th of a second. Since 1950 nearly all computers have been electronic, although the valve was superseded after a few years first by the transistor and more recently by what are called *integrated circuits*.

In the summer of 1946 two events of cardinal importance took place in Philadelphia. The first was the completion of ENIAC; the second the delivery of a course of lectures on 'The Theory and Techniques of Electronic Digital Compu-

FIGURE 120   The first electronic digital computer, the Electronic Numerical Integrator
and Calculator (ENIAC) (Smithsonian Institute, Washington DC)

ters'. We can now see, with the benefit of hindsight, that the 'logical design'
principles expounded in these lectures fixed the functional pattern of the electronic
computer as it is known today. The new ideas had been worked out by a group of
mathematicians and engineers led by the brilliant John von Neumann (1903–57).
They analysed the design problem in logical terms and then put forward proposals
for a computer which would be at once much smaller and far more powerful than
ENIAC. The construction of such a machine was put in hand at Philadelphia and a
number of other machines designed on the new principles soon appeared in various
parts of the world. The year 1946 thus marks the end of the 'computer middle-ages'
and the beginning of modern times.

British activity started in the winter of 1946/47, primarily at the Universities of
Manchester and Cambridge, and the National Physical Laboratory, near London.
During the next three years experimental computers were built at each of these
centres and served as prototypes for the commercially built computers which began
to appear in 1951: the Ferranti Mark I based on a Manchester prototype, the *Lyons
Electronic Office* (*LEO*) based on the *Electronic Delay Storage Automatic Computer* (*EDSAC*), the Cambridge computer; the English Electric *Digital Electronic
Universal Computing Engine* (*DEUCE*), based on the National Physical Laboratory prototype, the *Automatic Computing Engine* (*ACE*) *Pilot Model*.

In 1951 there were 15 to 20 computers more or less working in Britain; by 1959
they were about 200 (see examples in Figures 121 and 122). By the 1970s the
number had grown to some 6000 in Britain and more than 100 000 in the world as a
whole.

Thus in less than thirty years the electronic computer was able to penetrate into
every corner of society; today whole areas of public life are dependent on it. This
state of affairs is all the more remarkable when it is realised that a computer can

FIGURE 121   The Ferranti Mark I computer at Moston factory before delivery to Manchester University (courtesy Ferranti Ltd)

FIGURE 122   The GPO LEO 3 computer installation at Lytham St. Annes (courtesy the Post Office)

only do a very few simple things. Essentially, it works on patterns of symbols without distinguishing whether or not they are numbers. It compares these patterns and can determine whether or not they are identical and, if not, which comes first in an established ordered sequence of symbols. The computer is thus primarily a *logical* device rather than an *arithmetical* one. The logical aspects of computers can be described by a modified form of the algebra of George Boole (1815–1864), whilst the arithmetical aspects are based upon binary arithmetic (see below). In the case of numbers, it can move numbers around, form a third number from two numbers (for example by addition), and carry out simple logical ('yes' or 'no') tests. Why then is it such a powerful and ubiquitous tool? In elucidating a partial answer, three salient characteristics can be identified: *speed, versatility* and *power of choice*.

Speed   The immediate effect of applying electronic techniques to computing (in ENIAC) was an enormous increase in the speed of doing arithmetic. Computers are now much faster than ENIAC; by the 1970s it was possible to multiply two twelve-digit numbers together in a few millionths of a second – the time it takes a rifle bullet to travel a tenth of an inch, yet even this is slow by today's standards.

Versatility   The electronic computer is what is called a *digital* device; that is to say, it operates directly on numbers (or strictly speaking, on physical signals that a human being can interpret as numbers), just as we do when we calculate with pencil and paper. This means that it can carry out – at any rate in principle – any kind of calculation or symbol manipulation that can be broken down into a sequence of elementary arithmetical or logical steps: it is a *general-purpose* device. Lady Lovelace put the matter with her usual clarity when she wrote:

> The Analytical Engine has no pretension whatever to originate any thing. It can do whatever we *know how to order it to perform* [her italics].

In modern terminology, this means whatever we are able to express as an acceptable schedule of instructions, known as a *program* (the American spelling is now established usage). The first sentence of the quotation reminds us that the computing machine, unlike the human computer, is quite unable to make the smallest extension to its instructions when faced with an unforeseen situation: it will do exactly what it is told to do, no more and no less. The programmer must, therefore, foresee every contingency; he or she must provide precise instructions on what is to be done in any situation that might conceivably arise in the course of a calculation. (It is failure to do this that produces the kind of nonsense for which computers are notorious: the gas bill for £0.00, for example!) It is this combination of speed and versatility which makes the computer so fascinating. Not only can it do many diverse things, it can switch so rapidly from one to another that it seems to be doing a lot of different things at the same time.

Power of choice   By this is meant the ability of a computer to modify its own behaviour in the light of the current situation. It is obvious that the enormous speed of the electronic computer cannot be exploited if the machine has to be stopped after each step to enable the operator to make the settings for the next. The *high-speed* computer must of necessity be *automatic* in the sense already explained. It must therefore be provided in advance both with all the instructions (the *program*) and with all the *data* it needs for the complete job (or sequence of jobs). The term 'data' is here used to cover any information that is required; it may be numerical, alphabetical, logical or, more usually, a mixture of all three. Now this in turn demands a large *store* to hold instructions and data, and also any intermediate or final results that may be generated in the course of the calculation. Von Neumann argued in 1946 that it would be sensible to store everything together, and this is now the general practice (this explains the labels to be seen in Figure 123). Once common storage is accepted, it is clearly convenient to represent numbers and instructions inside the machine in the same way; in fact to *code* an instruction as a number, that is as a set of digits. Non-numerical data, such as words, are also coded as numbers inside the machine. A computer store can be thought of as consisting of a set of *registers*, each capable of being *addressed* individually in the program, and each holding either a number, an instruction coded as a number, or a group of letters also coded as a number. Nowadays the storage provided in a typical computer consists, in effect, of many millions of such registers, but each one is not a separate piece of equipment.

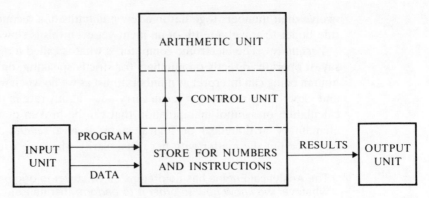

FIGURE 123   The five parts of an automatic computer

A typical computer can perform one million elementary operations every second; a typical calculation might take, say, two minutes. In this time the computer can execute more than one hundred million instructions! It is clearly impracticable either to write or to store a computer program of this length. The way out of the dilemma is to arrange matters so that a calculation is highly *repetitive*; so that groups of instructions are obeyed over and over again, but operating on different data each time. Fortunately most large computing jobs, especially in the commercial field, are like this anyway. Thus the same piece of program can be used for computing the pay of every person on the payroll, or for dealing with every enquiry (with perhaps a few exceptions) about booking a seat on an airline flight.

What is wanted is to be able to *cycle* repeatedly round a *loop* in the program until some 'signpost' is reached which tells us to leave the loop and do something else. It is convenient to make the choice of route depend on the answer to a simple 'yes or no' question such as: is the number in a specified register positive? The current value of this number would usually have been computed by the machine itself and would change from time to time during the course of the calculation. The programmer may well not know what path the calculation will take on a particular occasion; he or she prescribes the ground rules, as it were, and the machine applies them to the data presented to it. We can say, then, that the computer possesses a rudimentary power of choice, that it can take *decisions* as it goes along in the light of the current situation. However, we should always remember that it is the programmer who prescribes the rules of the game. Lady Lovelace knew all about this in 1842 when she wrote:

> The Engine is capable under certain circumstances of feeling about to discover which of two or more possible contingencies has occurred and then of shaping its future course accordingly.

She went further and discussed the mechanism whereby such a choice of route could be achieved. In the Jacquard loom the punched cards pass through the mechanism in a fixed order, which cannot be varied once the loom is set up. Lady Lovelace explained why an additional facility – which she calls 'backing' the cards – must be provided in the Analytical Engine. The drum over which the train of cards passes must be able to rotate in the reverse direction, the occasion and extent of the backward motion being determined by the program. She wrote:

> The object of this extension is to secure the possibility of bringing any particular card or set of cards in use *any number of times successively* in the solution of one problem [her italics].

A computer must be so designed that it can carry out a few basic operations when suitably stimulated to do so. More complex operations can be built up from these. The *basic instruction set* may be quite small, but it must contain at least one instruction (called variously the *test, discrimination, branching,* or *jump of control* instruction) which allows a choice between alternative routes. It is interesting to recall that the first working electronic stored-program computer in the world – a small laboratory model at the University of Manchester which first operated successfully in June 1948 – had a repertoire of only four instructions: transfer between store and arithmetic unit, subtract one number from another, a jump instruction of the kind we have been discussing, and stop. (Addition can be built up from two subtractions, but subtraction cannot be built up from two additions.)

Instructions are normally obeyed in the order in which they are stored in the registers. The choice is usually between obeying the next instruction in sequence or jumping to an instruction stored somewhere else. In accordance with the repetitive principle already mentioned, the jump is often back to some earlier part of the program. This is illustrated in Figure 124.

The effectiveness of a computer installation (the *hardware*) is totally dependent on the range and efficiency of its repertoire of programs (the *software*). For the last few decades, computer programming has been one of the fastest growing fields of

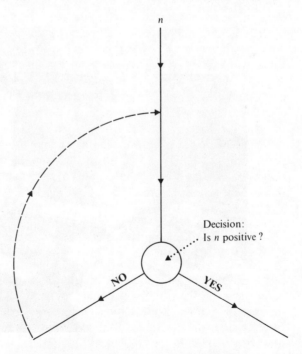

FIGURE 124   A decision point in the program

employment. It is not possible to pursue this very large subject here, but there is no shortage of books to assist the reader. Indeed, there are so many that it would be invidious to make a specific recommendation.

Modern programming

We have seen how the first digital computers were built in the laboratories of universities and research institutes. When the first production-built computers, most of them based on university models, began to appear in about 1952, they were mainly used to do the heavy numerical calculations that arise in some fields of scientific research, technical development and engineering design. Within a few years computers started to invade the business office. By the early 1960s commercial *data processing* (i.e. the fairly simple manipulation of large quantities of numerical and alphabetical information) had outstripped mathematical calculations as the main task of the growing army of computers.

This shift in the pattern of usage was made possible by a breakthrough in the technique of programming which took place between 1955 and 1960. The effect was that programs could be written in plain language, using simple English words with a minimum of mathematical symbolism. The basic ideas behind the breakthrough were very simple. The computer can *process* (i.e. manipulate and alter) numbers in its arithmetic unit. Program instructions are coded and stored as numbers, so the computer can be programmed to process instructions – if necessary, before they have ever been executed. It should therefore be possible to write a program for a particular job in one form – a form convenient to the programmer – and then to have the computer process (translate) it into another form – a form understandable

FIGURE 125   IBM 2250 display screen connected to an IBM 1130 computer (IBM)

by the machine. The translated program can then be executed. Therefore we need not address a computer in *its* language; we can teach it to understand *our* language.

In effect, a modern computer can be regarded as comprising vast numbers of minute interconnected electronic switches, each switch being in one of two given states at any particular moment. These states can be represented by 1 (usually interpreted as ON) and 0 (usually interpreted as OFF). This is analogous to verbal statements being either TRUE or FALSE, and can thus be related to logic. In 1847 the English mathematician George Boole, at that time a schoolmaster in Lincoln, had published a work with the title *The Mathematical Analysis of Logic*, which he later followed (in 1854, after having being appointed Professor of Mathematics at Queen's College, Cork) with *An Investigation into the Laws of Thought*. In these two works, Boole presented an algebra of logical operations which was to be used nearly a century later, though in a slightly revised form, for the logical design of computer switching circuits.

The switching circuits which perform arithmetic operations within computers are actually made up of combinations of logic circuits, which determine the state of any switch according to logical rather than arithmetic rules. We shall not go into the technical details of such circuits here; the important point to note is that the two digits 1 and 0 which can represent TRUE and FALSE in the logic circuits are precisely the two digits required for the binary number system (see p. 17). If therefore there is a sequence of switches, some of which may be ON and some OFF, this sequence can be interpreted as a binary number. For example such switches could be in the configuration:

$$\text{ON, OFF, OFF, ON, ON, OFF, OFF.}$$

Replacing ON and OFF by 1 and 0, respectively, yields 1001100, the binary representation of the decimal number 76 (i.e. 64 + 8 + 4). Computers, as configurations of switches, thus do their arithmetical calculations using binary numbers, though most users are not immediately aware of this since they work at keyboards and observe displays where numbers are represented decimally. It is probable that even fewer users are aware that it is logical algebra rather than arithmetic which forms the basis for the operations of the computer.

It is important not to confuse the logical algebra, using 1 and 0, with binary arithmetic. To illustrate a fundamental difference, consider the 'sum' of 1 and 1 in the two systems. In binary arithmetic, it is simply

$$1 + 1 = 10 \text{ (decimal 2)},$$

but in logical algebra

$$1 + 1 = 1.$$

This is because, in Logic, '+' can be interpreted as *or*. If a compound statement is made up of two simple statements joined by the conjunction *or*, and if the two individual simple statements are both TRUE, it necessarily follows that the whole statement is TRUE. Thus, the whole statement

'This paper is white *or* the ink is black

is TRUE, if it is TRUE that the paper is white and TRUE that the ink is black. Thus TRUE *or* TRUE yields TRUE. However, it is readily apparent that the whole statement is also

TRUE if just one of its parts is TRUE. This can be represented by

$$1 + 0 = 1 \text{ or } 0 + 1 = 1.$$

Unlike the first case, these two logical 'sums' are identical to the corresponding sums in binary arithmetic, as also is the expression

$$0 + 0 = 0,$$

(FALSE *or* FALSE yields FALSE). In the case of the logical 'product', where '.' may be interpreted as *and*, the logical and binary arithmetical expressions are identical, namely,

$$1 \cdot 1 = 1 \; ; \; 1 \cdot 0 = 0 \; ; \; 0 \cdot 1 = 0 \; ; \; 0 \cdot 0 = 0$$

The computer's logic circuits manipulate binary expressions, digit by digit, according to these simple logical rules. It is by appropriate and sometimes intricate combination of the logic circuits that computers perform arithmetic operations of extreme complexity.

Conclusions

This brief comparison of logical and binary arithmetic brings to an end the discussion of computers. The main theme throughout the book is the history of our decimal number system. The other topics, such as the various counting systems, tallying, fractions, the abacus, computers, and so on, have served principally to put that history in perspective.

Chapter 5 saw the final completion of the Hindu-Arabic decimal system, discussed in Chapter 4, with its extension to include fractions in the sixteenth century. However, this is a system for *writing* numbers invented in India as recently as the sixth century AD. The decimal *counting* system, which has an entirely separate history of its own, is much older, having been invented several millennia BC by the people who created the original Indo-European language, as shown in Chapter 3. Who exactly these people were is not known.

Equally unknown are the people or peoples who invented counting itself, where this occurred, and the exact nature of the cultural stress which brought it about. This information lies in the realms of pre-history, and the best that we can do is to speculate intelligently from such evidence as can be gathered. The *Seidenberg hypothesis* of single invention and diffusion of the various counting systems, discussed in Chapter 2, is supported by many facts but, as the writer admits, the evidence is not entirely conclusive. Even more speculative is the hypothesis of the ritual origin of counting put forward by the same writer. This last hypothesis has not been discussed here because it has no direct bearing on the book's principal theme of the history of our own decimal system. This theme has been explored using historical facts supported by documentary evidence, not speculation!

The gap between the primitive counting methods with which the book begins and the computers with which it ends may seem an enormous one. There are indeed great gaps both of concept and of sophistication. Nevertheless, running through the whole story has been the continuous thread of human desire to express the phenomena of the world in terms of numbers. Even the most sophisticated of modern computers needs numbers, though, perhaps surprisingly, not the ten digits of the decimal system but just the two digits of the binary system. Thus, in one sense, the arithmetic of computers can be seen as using a very 'primitive' system of numbers indeed, one that is even simpler than the earliest known system of counting, counting by pairs, which requires only the number words 'one' and 'two'. Had it had a representational numeral system corresponding to the binary and decimal systems, it would have required three symbols, not two.

There is no way, it would seem, that we can avoid dependence upon numbers; it is therefore essential for everyone to come to terms with them. If this book has provided an opportunity to obtain a deeper appreciation of numbers through its historical approach, it will, to a great extent, have fulfilled the objectives which the original authors of its content, including the present Editor, had in mind.

References and notes

1 Introduction

1. See especially K. Menninger (*trans.* by P. Broneer) *Number Words and Number Symbols* (MIT Press, 1969) and G. Flegg, *Numbers: their History and Meaning* (Deutsch and Schocken, 1983; Pelican Books, 1986).
2. A. Seidenberg, 'The Diffusion of Counting Practices', *University of California Publications in Mathematics* **3,** 4 (1960).

2 Counting systems

1. H Vedder, *Die Bergdama* (Hamburg, 1923) p. 165.
2. A. Seidenberg, 'The Diffusion of Counting Practices', *University of California Publications in Mathematics* **3,** 4 (1960), 216.
3. S. S. Dorman, *Pygmies and Bushmen of the Kalahari* (London, 1925) p. 61.
4. K. von den Steinen, *Unter den Naturvölkern Zentral Brasiliens* (Berlin, 1894) pp. 406–7, (translated by Margot McIntosh).
5. Seidenberg, *op. cit.*, p. 218.
6. *Ibid.*, pp. 218–9.
7. From Theodor Kluge, *Zahlenbegriffe der Völker Amerikas, Nordeurasiens, der Munda, und der Palaioafrikaner* (Berlin, 1939) p. 223.
8. K. Menninger, *Zahlwort und Ziffer* (1958) (*trans.* P. Broneer) *Number Words and Number Symbols* (MIT Press, 1969) p. 61.
9. *Ibid.*, p. 69.
10. O. Neugebauer, *The Exact Sciences in Antiquity* (Brown University Press, 1957).
11. Menninger, *op. cit.*, p. 49.
12. Seidenberg, *op. cit.*, p. 225.
13. Menninger, *op. cit.*, p. 36.
14. Seidenberg, *op. cit.*, p. 263.
15. Menninger, *op. cit.*, p. 36.
16. Seidenberg, *op. cit.*, p. 270.
17. Menninger, *op. cit.*, p. 223.
18. Taken from science-history journal *Isis*, xxvii, 462–3.
19. From C. Zaslavsky, *Africa Counts* (Prindle, Weber, and Schmidt, 1973) p. 18.
20. Menninger, *op. cit.*, p. 224.
21. *Ibid.*, p. 223.
22. See p. 8.
23. Menninger, *op. cit.*, p. 39.
24. From Menninger, *op. cit.*, p. 226.
25. G. Trevelyan, *History of England*, 2nd edn (Longman, Green, 1942) p. 125.
26. From which the expression 'bank stock'.
27. Menninger, *op. cit.*, p. 237.
28. From C. R. Jossett, *Money in Britain* (Warne, 1962) p. 17.
29. *Ibid.*, pp. 91–2.
30. From Menninger, *op. cit.*, p. 244.
31. *Ibid.*, pp. 253–4.
32. *Ibid.*, pp. 254.
33. Zasvlasky, *op. cit.*, pp. 93–5.

34. From the Latin *sextarius*, a 'sixth' (of the Roman measure of volume, the *congius*).
35. Menninger, *op. cit.*, p. 256.
36. A. Seidenberg, 'The Ritual Origin of Counting', *Archive for History of Exact Sciences* **2**, 1–40.

3 Number words

1. See C. B. Boyer, *A History of Mathematics* (Wiley, 1968) p. 683.
2. Raetia is now the canton of Grisons in Switzerland and part of the Tyrol.
3. That is, their endings are changed according to case or gender.
4. A word which is the subject of a verb is said to be in the *nominative* case, e.g. **I** give. The *accusative* denotes the direct object of a verb, e.g. I give **a coat**. The *genitive* denotes possession, e.g. I give my **father's** coat. The *dative* denotes the indirective (more remote) object, e.g. I give my father's coat to **him**.
5. K. Menninger, *Zahlwort und Ziffer* (1958) (*trans.* by P. Broneer) *Number Words and Number Symbols* (MIT Press, 1969) p. 22.
6. *Ibid.*, p. 106.
7. Pronounced as in German *acht* or Scottish *loch*.
8. Gothic *h = ch* as in *acht* or *loch*.
9. Gothic *ai* was probably pronounced rather like a short *e*.
10. From Datta and Singh, *History of Hindu Mathematics*, vol. 1 (Asia Publishing House 1936) pp. 10–11.
11. *Ibid.*, p. 12.
12. *Ibid.*, p. 13.
13. See, for example, Menninger, *op. cit.*, pp. 153–4.

4 Written numbers

1. All but a few fragments of the Rhind Papyrus are in the British Museum. It was published with commentary by T. E. Peet, London, 1923, and again by A. B. Chase, Ohio, 1929.
2. The numerals of Mesopotamia are discussed later.
3. From C. B. Boyer, 'Fundamental Steps in the Development of Numeration', *Isis* **35**, 157–8.
4. Babylonian fractions and methods of calculating are discussed further in Chapter 5.
5. The sun's yearly path.
6. Though, of course, using Greek numerals and not Babylonian wedges.
7. J. Needham, *Science and Civilization in China*, vol. 3 (Cambridge University Press, 1970) p. 10.
8. *Ibid.*, p. 12.
9. Boyer, *op. cit.*, pp. 761–8.
10. O. Neugebauer, 'Babylonian Mathematics', *Scripta Mathematica* **II,** 312–5.
11. See Chapter 2, p. 17.
12. Quotations and opinions criticised by Boyer are found in the following works:
    Cajori, F. *A History of Mathematical Notations* (Open Court, 1928–9)
    Cantor, M. *Varlesungen über Geschicte der Mathematik* (Leipzig, 1894)
    Crowther, J. G. *The Social Relations of Science* (New York, 1941)
    Delambre, J. B. J. *Histoire de l'astronomie ancienne* (Paris, 1917)
    Halsted, G. B. *On the Foundation and Technic of Arithmetic* (Chicago, 1912)
    Heath, T. L. *A History of Greek Mathematics* (Oxford, 1921)
    Hogben, L. *Mathematics for the Million* (New York, 1940)
    Karpinski, L. C. *History of Arithmetic* (Chicago and New York, 1925)
    Menninger, K. *Zahlwort und Ziffer* (1958), *trans.* P. Broneer, *Number Words and Number Symbols* (MIT Press, 1969)
    Tannery, P. *Mémoires,* J. L. Heiberg and H. G. Zeuthen (eds) (Toulouse and Paris, 1912–37).
13. See Note 12.

14. See Note 12.

15. Fractions are discussed in Chapter 5.

16. Decimal fractions are discussed in Chapter 5.

17. See Note 12.

18. See Note 12.

19. See pp. 102–30.

20. See p. 106.

21. For a fuller discussion of alphabetical numerals see: G. Flegg, *Numbers: their History and Meaning* (Deutsch and Schocken, 1983; Pelican Books, 1986) Chapter 3.

22. K. Menninger, *Zahlwort und Ziffer* (1958) (*trans.* P. Broneer) *Number Words and Number Symbols* (MIT Press, 1969) pp. 404–5.

23. The subject may be pursued further in B. Datta and A. N. Singh, *History of Hindu Mathematics* (Asia Publishing House, 1935–8).

24. For further details see O. Neugebauer and D. Pingree, *The Pañcasiddhântikâ of Varâhamihira* (Copenhagen, 1970).

25. This can be followed up in W. E. Clark, *The Âryabhatîya of Âryabhata* (Chicago, 1930).

26. Datta and Singh, *op. cit.*, p. 69.

27. Described in Datta and Singh, *op. cit.*, pp. 69–72.

28. In a right-angled triangle with hypotenuse, the *sine* of an angle is equal to the length of the side opposite to it.

29. Datta and Singh (*op. cit.*, p. 58, footnote 5) describe Pargiter as 'probably the greatest Puranic scholar of modern times'.

30. It is translated with an excellent commmentary by Burgess and Whitney in *Journal of the American Oriental Society* **6** (1960) 141–498.

31. See Datta and Singh, *op. cit.*, p. 40.

32. Extensive material on Arabic forms of numerals, with many examples from manuscripts, may be found in Rida A. K. Irani, 'Arabic Numeral Forms', *Centaurus* **4** (1955) 1–12.

33. Planudes (1255?–1310) was a Greek monk, Ambassador to Venice of the Emperor Andronicus II. He wrote a work on Hindu numerals and also a commentary on Diophantus' *Arithmetic*.

34. Bibliothèque Nationale, Paris, *lat.* 8663, *f.* 49.

35. The orientations of the numerals are discussed in detail in G. Beaujouan 'Étude paléographique sur la rotation des chiffres et l'emploi des apices du Xe au Xii^e siècle', *Revue d'histoire des sciences* **1**, 301–13.

36. Neither Pythagoras nor Boethius is a suitable historical choice.

5 Fractions and calculation

1. It was published with an excellent commentary by T. E. Peet (Liverpool and London, 1923, reprinted by Kraus Reprint, Nendeln, Lichtenstein, 1970).

2. See pp. 78–80 for a comment on the relative importance of hieroglyphic and hieratic script.

3. This hypothesis was put forward by O. Neugebauer in his pioneer paper in *Quellen und Studien zur Geschichte der Math., Astron. und Physik* **2** (1930) 301.

4. Published by Schack-Schackenburg in *Aegypt. Zeitsch* **38** and **40** (1900).

5. Yale Babylonian Collection (YBC) 10529. The text is reproduced in O. Neugebauer and A. Sachs, *Mathematical Cuneiform Texts* (Newhaven, Connecticut, 1945) p. 16.

6. See B. L. van der Waerden, *Science Awakening I* (Noordhoff) pp. 97–100.

7. *Ibid.*, p. 66 for fuller discussion.

8. *Ibid.*, pp. 70–1.

9. This explanation is a 'possible' one only. Some scholars (e.g. Boyer and Neugebauer) specifically reject it.

10. In a right-angled triangle with one further angle given, the ratio of the sides is determined. The ratio of the side opposite to a given angle to the hypotenuse is called the sine of the angle. The sine of 30° is $\frac{1}{2}$ hence, in this example $\frac{e}{g} = \frac{1}{2}$ and if $e = 1$ then $g = 2$.

11. K. Vogel, *Griechische Logistik*, Sitzungsber. Bayer Akad. Munchen (Math.-Nat.) p. 357.
12. Manchu dynasty, 1644–1911.
13. For a discussion of the Euclidean algorithm see, for example, C. B. Boyer, *A History of Mathematics* (Wiley, 1968) pp. 126–7.
14. From Datta and Singh, *History of Hindu Mathematics*, Part 1 (Asia Publishing House, 1962) p. 189.
15. *Ibid.*, p. 190.
16. *Ibid.*, p. 196.
17. *Ibid.*, p. 196.
18. *Ibid.*, p. 198.
19. *Ibid.*, p. 188.
20. *Ibid.*, p. 170.
21. For example, *ibid.*, pp. 155–60.
22. *Ibid.*, p. 160.
23. *Ibid.*, p. 161.
24. There is an English translation with extensive commentary by Martin Luther d'Ooge (New York, 1926, reprinted, 1972).
25. From Datta and Singh, *op. cit.*, pp. 291–20.
26. See P. Luckey, *Der Lehrbrief uber Kreisumfrang von Gamsid b. Mas'ud al-Kashi* (Berlin, 1953).
27. See A. P. Juschkowitsch, *Mathematik im Mittelalter* (Leipzig, 1964) p. 241.
28. Norton's English translation was reprinted, together with a facsimile of the original Dutch edition, in volume 2 of the *Principal Works of Simon Stevin*, D. J. Struik (ed.) (Amsterdam, 1958).

6 Aids to Calculation

1. K. Menninger, *Zahlwort und Ziffer* (1958) (*trans.* by P. Broneer) *Number Words and Number Symbols* (MIT Press, 1969) pp. 301–2.
2. *Artes* here means 'practical skills'.
3. See J. Needham, *Science and Civilization in China*, Vol. 3 (Cambridge University Press, 1970) p. 8.
4. *Ibid.*, p. 70.

Bibliography

The following list includes histories of mathematics and source books which are commended for those wishing to read more widely in the subject.

Boyer, C.B. *A History of Mathematics* (Wiley, 1968).

Dedron, P. and Itard, J. (*trans.* Field, J.) *Mathematics and Mathematicians* (2 vols.) (Transworld/Open University Press, 1974).

Eves, H. *An Introduction to the History of Mathematics* (Holt, 1964).

Fauvel, J. and Gray, J. (eds) *The History of Mathematics: a Reader* (Macmillan, 1987).

Flegg, G. (1983) *Numbers: their History and Meaning* (Deutsch and Schocken/Pelican, 1983).

Flegg, G. *Boolean Algebra* (Macdonald/Transworld, 1971/2).

Hollingdale, S.H. and Toothill, C. *Electronic Computers* (Penguin, 1975).

Karpinski, L.C. *The History of Arithmetic* (Russell and Russell, 1965).

Kline, M. *Mathematical Thought from Ancient to Modern Times* (Oxford, 1972).

Kline, M. *Mathematics in Western Culture* (Penguin, 1972).

Lavington, S. *Early British Computers* (Manchester University Press, 1980).

Martin, J. and Normand, A.R.D. *The Computerised Society* (Penguin, 1973).

Menninger, K. (*trans.* Broneer, P.) *Number Words and Number Symbols* (MIT Press, 1970).

Morrison, P. and E. (eds) *Charles Babbage and his Calculating Engines* (Dover, 1961).

Midonick, H. *The Treasury of Mathematics* (2 vols) (Penguin, 1968).

Neugebauer, O. *The Exact Sciences in Antiquity* (Dover, 1969).

Newman, J. *The World of Mathematics* (4 vols) (Allen and Unwin, 1960).

Popp, W. *trans.* Bruckheimer, M. *History of Mathematics: Topics for Schools* (Transworld/Open University Press, 1975).

Smith, D.E. *A Source Book in Mathematics* (2 vols) (Dover, 1959).

Struik, D.J. *A Concise History of Mathematics* (Bell, 1954).

Struik, D.J. *A Source Book in Mathematics* (Harvard University Press, 1969).

Van der Waerden, B.L. *Science Awakening I* (Oxford, 1961).

Wilder, R.L. *Evolution of Mathematical Concepts* (Transworld/Open University Press, 1974).

Index